Mathematics of Data

Volume 1

How to reveal, characterize, and exploit the structure in data? Meeting this central challenge of modern data science requires the development of new mathematical approaches to data analysis, going beyond traditional statistical methods. Fruitful mathematical methods can originate in geometry, topology, algebra, analysis, stochastics, combinatorics, or indeed virtually any field of mathematics. Confronting the challenge of structure in data is already leading to productive new interactions among mathematics, statistics, and computer science, notably in machine learning. We invite novel contributions (research monographs, advanced textbooks, and lecture notes) presenting substantial mathematics that is relevant for data science. Since the methods required to understand data depend on the source and type of the data, we very much welcome contributions comprising significant discussions of the problems presented by particular applications. We also encourage the use of online resources for exercises, software and data sets. Contributions from all mathematical communities that analyze structures in data are welcome. Examples of potential topics include optimization, topological data analysis, compressed sensing, algebraic statistics, information geometry, manifold learning, tensor decomposition, support vector machines, neural networks, and many more.

Hal Schenck

Algebraic Foundations
for Applied Topology
and Data Analysis

 Springer

Hal Schenck
Department of Mathematics and Statistics
Auburn University
Auburn, AL, USA

ISSN 2731-4103   ISSN 2731-4111   (electronic)
Mathematics of Data
ISBN 978-3-031-06666-5   ISBN 978-3-031-06664-1   (eBook)
https://doi.org/10.1007/978-3-031-06664-1

This work was supported by National Science Foundation (2006410) (http://dx.doi.org/10.13039/100000001) and Rosemary Kopel Brown Endowment

Mathematics Subject Classification: 55N31, 18G99

This Springer imprint is published by the registered company Springer Nature Switzerland AG
The registered company address is: Gewerbestrasse 11, 6330 Cham, Switzerland

Preface

This book is a mirror of applied topology and data analysis: it covers a wide range of topics, at levels of sophistication varying from the elementary (matrix algebra) to the esoteric (Grothendieck spectral sequence). My hope is that there is something for everyone, from undergraduates immersed in a first linear algebra class to sophisticates investigating higher dimensional analogs of the barcode. Readers are encouraged to barhop; the goal is to give an intuitive and hands-on introduction to the topics, rather than a punctiliously precise presentation.

The notes grew out of a class taught to statistics graduate students at Auburn University during the COVID summer of 2020. The book reflects that: it is written for a mathematically engaged audience interested in understanding the theoretical underpinnings of topological data analysis. Because the field draws practitioners with a broad range of experience, the book assumes little background at the outset. However, the advanced topics in the latter part of the book require a willingness to tackle technically difficult material.

To treat the algebraic foundations of topological data analysis, we need to introduce a fair number of concepts and techniques, so to keep from bogging down, proofs are sometimes sketched or omitted. There are many excellent texts on upper-level algebra and topology where additional details can be found, for example:

Algebra References	Topology References
Aluffi [2]	Fulton [76]
Artin [3]	Greenberg–Harper [83]
Eisenbud [70]	Hatcher [91]
Hungerford [93]	Munkres [119]
Lang [102]	Spanier [137]
Rotman [132]	Weibel [149]

Techniques from linear algebra have been essential tools in data analysis from the birth of the field, and so the book kicks off with the basics:

- Least Squares Approximation
- Covariance Matrix and Spread of Data
- Singular Value Decomposition

Tools from topology have recently made an impact on data analysis. This text provides the background to understand developments such as *persistent homology*. Suppose we are given point cloud data, that is, a set of points X:

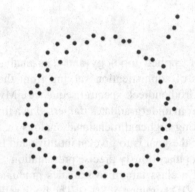

If X was sampled from some object Y, we'd like to use X to infer properties of Y. Persistent homology applies tools of algebraic topology to do this. We start by using X as a *seed* from which to *grow* a family of spaces

$$X_\epsilon = \bigcup_{p \in X} N_\epsilon(p), \text{ where } N_\epsilon(p) \text{ denotes an } \epsilon \text{ ball around } p.$$

As $X_\epsilon \subseteq X_{\epsilon'}$ if $\epsilon \leq \epsilon'$, we have a family of topological spaces and inclusion maps.

As Weinberger notes in [150], persistent homology is a type of Morse theory: there are a finite number of values of ϵ where the topology of X_ϵ changes. Notice that when $\epsilon \gg 0$, X_ϵ is a giant blob; so ϵ is typically restricted to a range $[0, x]$. Topological features which "survive" to the parameter value x are said to be *persistent*; in the example above, the circle S^1 is a persistent feature.

The first three chapters of the book are an algebra-topology boot camp. Chapter 1 provides a brisk review of the tools from linear algebra most relevant for applications, such as webpage ranking. Chapters 2 and 3 cover the results we need from upper-level classes in (respectively) algebra and topology. Applied topology appears in Sect. 3.4, which ties together sheaves, the heat equation, and social media. Some readers may want to skip the first three chapters, and jump in at Chap. 4.

The main techniques appear in Chap. 4, which defines simplicial complexes, simplicial homology, and the Čech & Rips complexes. These concepts are illustrated with a description of the work of de Silva–Ghrist using the Rips complex to analyze sensor networks. Chapter 5 further develops the algebraic topology toolkit, introducing several cohomology theories and highlighting the work of Jiang–Lim– Yao–Ye in applying Hodge theory to voter ranking (the Netflix problem).

We return to algebra in Chap. 6, which is devoted to modules over a principal ideal domain—the structure theorem for modules over a principal ideal domain plays a central role in persistent homology. Chapters 7 and 8 cover, respectively, persistent and multiparameter persistent homology. Chapter 9 is the pièce de résistance (or perhaps the coup de grâce)—a quick and dirty guide to derived functors and spectral sequences. Appendix A illustrates several of the software packages which can be used to perform computations.

There are a number of texts which tackle data analysis from the perspective of pure mathematics. *Elementary Applied Topology* [79] by Rob Ghrist is closest in spirit to these notes, it has a more topological flavor (and wonderfully illuminating illustrations!). Other texts with a similar slant include *Topological Data Analysis with Applications* [30] by Carlsson–Vejdemo-Johanssen, *Computational Topology for Data Analysis* [60] by Dey–Wang, and *Computational Topology* [67] by Edelsbrunner–Harer. At the other end of the spectrum, *Persistence Theory: From Quiver Representations to Data Analysis* [124] by Steve Oudot is intended for more advanced readers; the books of Polterovich–Rosen–Samvelyan–Zhang [126], Rabadan–Blumberg [127], and Robinson [131] focus on applications. Statistical methods are not treated here; they merit a separate volume.

Many folks deserve acknowledgment for their contributions to these notes, starting with my collaborators in the area: Heather Harrington, Nina Otter, and Ulrike Tillmann. A precursor to the summer TDA class was a minicourse I taught at CIMAT, organized by Abraham Martín del Campo, Antonio Rieser, and Carlos Vargas Obieta. The summer TDA class was itself made possible by the Rosemary Kopel Brown endowment, and writing by NSF grant 2048906. Thanks to Hannah Alpert, Ulrich Bauer, Michael Catanzaro, David Cox, Ketai Chen, Carl de Boor, Vin de Silva, Robert Dixon, Jordan Eckert, Ziqin Feng, Oliver Gäfvert, Rob Ghrist, Sean Grate, Anil Hirani, Yang Hui He, Michael Lesnick, Lek-Heng Lim, Dmitriy Morozov, Steve Oudot, Alex Suciu, and Matthew Wright for useful feedback.

Data science is a moving target: one of today's cutting-edge tools may be relegated to the ash bin of history tomorrow. For this reason, the text aims to highlight mathematically important concepts which have *proven* or *potential* utility in applied topology and data analysis. But mathematics is a human endeavor, so it is wise to remember Seneca: "Omnia humana brevia et caduca sunt."

Auburn, AL, USA   Hal Schenck
September 2022

Contents

Chapter 1
Linear Algebra Tools for Data Analysis

We begin with a short and intense review of linear algebra. There are a number of reasons for this approach; first and foremost is that linear algebra is the computational engine that drives most of mathematics, from numerical analysis (finite element method) to algebraic geometry (Hodge decomposition) to statistics (covariance matrix and the shape of data). A second reason is that practitioners of data science come from a wide variety of backgrounds—a statistician may not have seen eigenvalues since their undergraduate days. The goal of this chapter is to develop basic dexterity dealing with

- Linear Equations, Gaussian Elimination, Matrix Algebra.
- Vector Spaces, Linear Transformations, Basis and Change of Basis.
- Diagonalization, Webpage Ranking, Data and Covariance.
- Orthogonality, Least Squares Data Fitting, Singular Value Decomposition.

There are entire books written on linear algebra, so our focus will be on examples and computation, with proofs either sketched or left as exercises.

1.1 Linear Equations, Gaussian Elimination, Matrix Algebra

In this section, our goal is to figure out how to set up a system of equations to study the following question:

Example 1.1.1 Suppose each year in Smallville that 30% of nonsmokers begin to smoke, while 20% of smokers quit. If there are 8000 smokers and 2000 nonsmokers at time $t = 0$, after 100 years, what are the numbers? After n years? Is there an equilibrium state at which the numbers stabilize?

© The Author(s), under exclusive license to Springer Nature Switzerland AG 2022
H. Schenck, *Algebraic Foundations for Applied Topology and Data Analysis*,
Mathematics of Data 1, https://doi.org/10.1007/978-3-031-06664-1_1

The field of *linear algebra* is devoted to analyzing questions of this type. We now embark on a quick review of the basics. For example, consider the system of linear equations

$$x - 2y = 2$$
$$2x + 3y = 6$$

Gaussian elimination provides a systematic way to manipulate the equations to reduce to fewer equations in fewer variables. A linear equation (no variable appears to a power greater than one) in one variable is of the form $x =$ constant, so is trivial. For the system above, we can eliminate the variable x by multiplying the first row by -2 (note that this does not change the solutions to the system), and adding it to the second row, yielding the new equation $7y = 2$. Hence $y = \frac{2}{7}$, and substituting this for y in either of the two original equations, we solve for x, finding that $x = \frac{18}{7}$. Gaussian elimination is a formalization of this simple example: given a system of linear equations

$$a_{11}x_1 + a_{12}x_2 + \cdots a_{1n}x_n = b_1$$
$$a_{21}x_1 + a_{22}x_2 + \cdots a_{2n}x_n = b_2$$
$$\vdots \qquad\qquad \vdots \quad \vdots$$

(a) swap order of equations so $a_{11} \neq 0$.
(b) multiply the first row by $\frac{1}{a_{11}}$, so the coefficient of x_1 in the first row is 1.
(c) subtract a_{i1} times the first row from row i, for all $i \geq 2$.

At the end of this process, only the first row has an equation with nonzero x_1 coefficient. Hence, we have reduced to solving a system of fewer equations in fewer unknowns. Iterate the process. It is important that the operations above do not change the solutions to the system of equations; they are known as *elementary row operations*: formally, these operations are

(a) Interchange two rows.
(b) Multiply a row by a nonzero constant.
(c) Add a multiple of one row to another row.

Exercise 1.1.2 Solve the system

$$x + y + z = 3$$
$$2x + y = 7$$
$$3x + 2z = 5$$

You should end up with $z = -\frac{7}{5}$, now backsolve. It is worth thinking about the geometry of the solution set. Each of the three equations defines a plane in $\mathbb{R}^3$.

What are the possible solutions to the system? If two distinct planes are parallel, we have equations $ax + by + cz = d$ and $ax + by + cz = e$, with $d \neq e$, then there are no solutions, since no point can lie on both planes. On the other hand, the three planes could meet in a single point—this occurs for the system $x = y = z = 0$, for which the origin $(0, 0, 0)$ is the only solution. Other geometric possibilities for the set of common solutions are a line or plane; describe the algebra that corresponds to the last two possibilities. $\diamond$

There is a simple shorthand for writing a system of linear equations as above using matrix notation. To do so, we need to define matrix multiplication.

Definition 1.1.3 A matrix is a $m \times n$ array of elements, where m is the number of rows and n is the number of columns.

Vectors are defined formally in §1.2; informally we think of a real vector $\mathbf{v}$ as an $n \times 1$ or $1 \times n$ matrix with entries in $\mathbb{R}$, and visualize it as a directed arrow with tail at the origin $(0, \ldots, 0)$ and head at the point of $\mathbb{R}^n$ corresponding to $\mathbf{v}$.

Definition 1.1.4 The dot product of vectors $\mathbf{v} = [v_1, \ldots, v_n]$ and $\mathbf{w} = [w_1, \ldots, w_n]$ is

$$\mathbf{v} \cdot \mathbf{w} = \sum_{i=1}^{n} v_i w_i, \text{ and the length of } \mathbf{v} \text{ is } |\mathbf{v}| = \sqrt{\mathbf{v} \cdot \mathbf{v}}.$$

By the law of cosines, $\mathbf{v}, \mathbf{w}$ are orthogonal iff $\mathbf{v} \cdot \mathbf{w} = 0$, which you'll prove in Exercise 1.4.1. An $m \times n$ matrix A and $p \times q$ matrix B can be multiplied when $n = p$. If $(AB)_{ij}$ denotes the (i, j) entry in the product matrix, then

$$(AB)_{ij} = \text{row}_i(A) \cdot \text{col}_j(B).$$

This definition may seem opaque. It is set up *exactly* so that when the matrix B represents a transition from State$_1$ to State$_2$ and the matrix A represents a transition from State$_2$ to State$_3$, then the matrix AB represents the composite transition from State$_1$ to State$_3$. This makes clear the reason that the number of columns of A must be equal to the number of rows of B to compose the operations: the *target* of the map B is the *source* of the map A.

Exercise 1.1.5

$$\begin{bmatrix} 2 & 7 \\ 3 & 3 \\ 1 & 5 \end{bmatrix} \cdot \begin{bmatrix} 1 & 2 & 3 & 4 \\ 5 & 6 & 7 & 8 \end{bmatrix} = \begin{bmatrix} 37 & 46 & 55 & 64 \\ * & * & * & * \\ * & * & * & * \end{bmatrix}$$

Fill in the remainder of the entries. $\diamond$

Definition 1.1.6 The transpose A^T of a matrix A is defined via $(A^T)_{ij} = A_{ji}$. A is symmetric if $A^T = A$, and diagonal if $A_{ij} \neq 0 \Rightarrow i = j$. If A and B are diagonal $n \times n$ matrices, then

$$AB = BA \text{ and } (AB)_{ii} = a_{ii} \cdot b_{ii}$$

The $n \times n$ identity matrix I_n is a diagonal matrix with 1's on the diagonal; if A is an $n \times m$ matrix, then

$$I_n \cdot A = A = A \cdot I_m.$$

An $n \times n$ matrix A is invertible if there is an $n \times n$ matrix B such that $BA = AB = I_n$. We write A^{-1} for the matrix B; the matrix A^{-1} is the *inverse* of A.

Exercise 1.1.7 Find a pair of 2×2 matrices (A, B) such that $AB \neq BA$. Show that $(AB)^T = B^T A^T$, and use this to prove that the matrix $A^T A$ is symmetric.   ◇

In matrix notation a system of n linear equations in m unknowns is written as

$$A \cdot \mathbf{x} = \mathbf{b}, \text{ where } A \text{ is an } n \times m \text{ matrix and } \mathbf{x} = [x_1, \ldots, x_m]^T.$$

We close this section by returning to our vignette.

Example 1.1.8 To start the analysis of smoking in Smallville, we write out the matrix equation representing the change during the first year, from $t = 0$ to $t = 1$. Let $[n(t), s(t)]^T$ be the vector representing the number of nonsmokers and smokers (respectively) at time t. Since 70% of nonsmokers continue as nonsmokers, and 20% of smokers quit, we have $n(1) = .7n(0) + .2s(0)$. At the same time, 30% of nonsmokers begin smoking, while 80% of smokers continue smoking, hence $s(1) = .3n(0) + .8s(0)$. We encode this compactly as the matrix equation:

$$\begin{bmatrix} n(1) \\ s(1) \end{bmatrix} = \begin{bmatrix} .7 & .2 \\ .3 & .8 \end{bmatrix} \cdot \begin{bmatrix} n(0) \\ s(0) \end{bmatrix}$$

Now note that to compute the smoking status at $t = 2$, we have

$$\begin{bmatrix} n(2) \\ s(2) \end{bmatrix} = \begin{bmatrix} .7 & .2 \\ .3 & .8 \end{bmatrix} \cdot \begin{bmatrix} n(1) \\ s(1) \end{bmatrix} = \begin{bmatrix} .7 & .2 \\ .3 & .8 \end{bmatrix} \cdot \begin{bmatrix} .7 & .2 \\ .3 & .8 \end{bmatrix} \cdot \begin{bmatrix} n(0) \\ s(0) \end{bmatrix}$$

And so on, ad infinitum. Hence, to understand the behavior of the system for t very large (written $t \gg 0$), we need to compute

$$\left(\lim_{t \to \infty} \begin{bmatrix} .7 & .2 \\ .3 & .8 \end{bmatrix}^t \right) \cdot \begin{bmatrix} n(0) \\ s(0) \end{bmatrix}$$

Matrix multiplication is computationally expensive, so we'd like to find a trick to save ourselves time, energy, and effort. The solution is the following, which for now will be a *Deus ex Machina* (but lovely nonetheless!) Let A denote the 2×2 matrix above (multiply by 10 for simplicity). Then

$$\begin{bmatrix} 3/5 & -2/5 \\ 1/5 & 1/5 \end{bmatrix} \cdot \begin{bmatrix} 7 & 2 \\ 3 & 8 \end{bmatrix} \cdot \begin{bmatrix} 1 & 2 \\ -1 & 3 \end{bmatrix} = \begin{bmatrix} 5 & 0 \\ 0 & 10 \end{bmatrix}$$

Write this equation as $BAB^{-1} = D$, with D denoting the diagonal matrix on the right hand side of the equation. An easy check shows BB^{-1} is the identity matrix I_2. Hence

$$(BAB^{-1})^n = (BAB^{-1})(BAB^{-1})(BAB^{-1}) \cdots BAB^{-1} = BA^n B^{-1} = D^n,$$

which follows from collapsing the consecutive $B^{-1}B$ terms in the expression. So

$$BA^n B^{-1} = D^n, \text{ and therefore } A^n = B^{-1}D^n B.$$

As we saw earlier, multiplying a diagonal matrix with itself costs nothing, so we have reduced a seemingly costly computation to *almost* nothing; how do we find the mystery matrix B? The next two sections of this chapter are devoted to answering this question.

1.2   Vector Spaces, Linear Transformations, Basis and Change of Basis

In this section, we lay out the underpinnings of linear algebra, beginning with the definition of a *vector space* over a *field* $\mathbb{K}$. A field is a type of *ring*, and is defined in detail in the next chapter. For our purposes, the field $\mathbb{K}$ will typically be one of $\{\mathbb{Q}, \mathbb{R}, \mathbb{C}, \mathbb{Z}/p\}$, where p is a prime number.

Definition 1.2.1 (Informal) A vector space V is a collection of objects (vectors), endowed with two operations: vectors can be added to produce another vector, or multiplied by an element of the field. Hence the set of vectors V is closed under the operations. The formal definition of a vector space appears in Definition 2.2.1.

Example 1.2.2 Examples of vector spaces.

(a) $V = \mathbb{K}^n$, with $[a_1, \ldots, a_n] + [b_1, \ldots, b_n] = [a_1 + b_1, \ldots, a_n + b_n]$ and $c[a_1, \ldots, a_n] = [ca_1, \ldots, ca_n]$.
(b) The set of polynomials of degree at most $n - 1$, with coefficients in $\mathbb{K}$. Show this has the same structure as part (a).
(c) The set of continuous functions on the unit interval.

If we think of vectors as arrows, then we can visualize vector addition as putting the tail of one vector at the head of another: draw a picture to convince yourself that

$$[1, 2] + [2, 4] = [3, 6].$$

1.2.1 Basis of a Vector Space

Definition 1.2.3 For a vector space V, a set of vectors $\{\mathbf{v}_1, \ldots, \mathbf{v}_k\} \subseteq V$ is

- linearly independent (or simply *independent*) if

$$\sum_{i=1}^{k} a_i \mathbf{v}_i = 0 \Rightarrow \text{ all the } a_i = 0.$$

- a spanning set for V (or simply *spans* V) if for any $\mathbf{v} \in V$ there exist $a_i \in \mathbb{K}$ such that

$$\sum_{i=1}^{k} a_i \mathbf{v}_i = \mathbf{v}.$$

Example 1.2.4 The set $\{[1, 0], [0, 1], [2, 3]\}$ is dependent, since $2 \cdot [1, 0] + 3 \cdot [0, 1] - 1 \cdot [2, 3] = [0, 0]$. It is a spanning set, since an arbitrary vector $[a, b] = a \cdot [1, 0] + b \cdot [0, 1]$. On the other hand, for $V = \mathbb{K}^3$, the set of vectors $\{[1, 0, 0], [0, 1, 0]\}$ is independent, but does not span.

Definition 1.2.5 A subset

$$S = \{\mathbf{v}_1, \ldots, \mathbf{v}_k\} \subseteq V$$

is a *basis* for V if it spans and is independent. If S is finite, we define the dimension of V to be the cardinality of S.

A basis is loosely analogous to a set of letters for a language where the words are vectors. The spanning condition says we have enough letters to write every word, while the independent condition says the representation of a word in terms of the set of letters is unique. The vector spaces we encounter in this book will be finite dimensional. A cautionary word: there are subtle points which arise when dealing with infinite dimensional vector spaces.

Exercise 1.2.6 Show that the dimension of a vector space is well defined.

1.2.2 Linear Transformations

One of the most important constructions in mathematics is that of a mapping between objects. Typically, one wants the objects to be of the same type, and for the map to preserve their structure. In the case of vector spaces, the right concept is that of a *linear transformation*:

Definition 1.2.7 Let V and W be vector spaces. A map $T : V \to W$ is a *linear transformation* if

$$T(c\mathbf{v}_1 + \mathbf{v}_2) = cT(\mathbf{v}_1) + T(\mathbf{v}_2),$$

for all $\mathbf{v}_i \in V$, $c \in \mathbb{K}$. Put more tersely, sums split up, and scalars pull out.

While a linear transformation may seem like an abstract concept, the notion of basis will let us represent a linear transformation via matrix multiplication. On the other hand, our vignette about smokers in Smallville in the previous section shows that not all representations are equal. This brings us to *change of basis*.

Example 1.2.8 The sets $B_1 = \{[1, 0], [1, 1]\}$ and $B_2 = \{[1, 1], [1, -1]\}$ are easily checked to be bases for $\mathbb{K}^2$. Write $\mathbf{v}_{B_i}$ for the representation of a vector $\mathbf{v}$ in terms of the basis B_i. For example

$$[0, 1]_{B_1} = 0 \cdot [1, 0] + 1 \cdot [1, 1] = 1 \cdot [1, 1] + 0 \cdot [1, -1] = [1, 0]_{B_2}$$

The algorithm to write a vector $\mathbf{b}$ in terms of a basis $B = \{\mathbf{v}_1, \ldots, \mathbf{v}_n\}$ is as follows: construct a matrix A whose columns are the vectors $\mathbf{v}_i$, then use Gaussian elimination to solve the system $A\mathbf{x} = \mathbf{b}$.

Exercise 1.2.9 Write the vector $[2, 1]$ in terms of the bases B_1 and B_2 above. ◇

To represent a linear transformation $T : V \to W$, we need to have frames of reference for the source and target–this means choosing bases $B_1 = \{\mathbf{v}_1, \ldots, \mathbf{v}_n\}$ for V and $B_2 = \{\mathbf{w}_1, \ldots, \mathbf{w}_m\}$ for W. Then the matrix $M_{B_2 B_1}$ representing T with respect to *input* in basis B_1 and *output* in basis B_2 has as i^{th} column the vector $T(\mathbf{v}_i)$, written with respect to the basis B_2. An example is in order:

Example 1.2.10 Let $V = \mathbb{R}^2$, and let T be the transformation that rotates a vector counterclockwise by 90 degrees. With respect to the standard basis $B_1 = \{[1, 0], [0, 1]\}$, $T([1, 0]) = [0, 1]$ and $T([0, 1]) = [-1, 0]$, so

$$M_{B_1 B_1} = \begin{bmatrix} 0 & -1 \\ 1 & 0 \end{bmatrix}$$

Using the bases B_1 (for input) and B_2 (for output) from Example 1.2.8 yields

$$
\begin{aligned}
T([1,0]) &= [0,1] &= 1/2 \cdot [1,1] - 1/2 \cdot [1,-1] \\
T([1,1]) &= [-1,1] &= 0 \cdot [1,1] - 1 \cdot [1,-1]
\end{aligned}
$$

So

$$
M_{B_2 B_1} = \begin{bmatrix} 1/2 & 0 \\ -1/2 & -1 \end{bmatrix}
$$

1.2.3 Change of Basis

Suppose we have a representation of a matrix or vector with respect to basis B_1, but need the representation with respect to basis B_2. This is analogous to a German speaking diner being presented with a menu in French: we need a *translator* (or language lessons!)

Definition 1.2.11 Let B_1 and B_2 be two bases for the vector space V. The change of basis matrix Δ_{21} takes as input a vector represented in basis B_1, and outputs the same vector represented with respect to the basis B_2.

Algorithm 1.2.12 *Given bases $B_1 = \{\mathbf{v}_1, \ldots, \mathbf{v}_n\}$ and $B_2 = \{\mathbf{w}_1, \ldots, \mathbf{w}_n\}$ for V, to find the change of basis matrix Δ_{21}, form the $n \times 2n$ matrix whose first n columns are B_2 (the "new" basis), and whose second n columns are B_1 (the "old" basis). Row reduce to get a matrix whose leftmost $n \times n$ block is the identity. The rightmost $n \times n$ block is Δ_{21}. The proof below for $n = 2$ generalizes easily.*

Proof Since B_2 is a basis, we can write

$$
\begin{aligned}
\mathbf{v}_1 &= \alpha \cdot \mathbf{w}_1 + \beta \cdot \mathbf{w}_2 \\
\mathbf{v}_2 &= \gamma \cdot \mathbf{w}_1 + \delta \cdot \mathbf{w}_2
\end{aligned}
$$

and therefore

$$
\begin{bmatrix} a \\ b \end{bmatrix}_{B_1} = a \cdot \mathbf{v}_1 + b \cdot \mathbf{v}_2 = a \cdot (\alpha \cdot \mathbf{w}_1 + \beta \cdot \mathbf{w}_2) + b \cdot (\gamma \cdot \mathbf{w}_1 + \delta \cdot \mathbf{w}_2) = \left(\begin{bmatrix} \alpha & \gamma \\ \beta & \delta \end{bmatrix} \cdot \begin{bmatrix} a \\ b \end{bmatrix} \right)_{B_2}
$$

$\square$

Example 1.2.13 Let $B_1 = \{[1, 0], [0, 1]\}$ and $B_2 = \{[1, 1], [1, -1]\}$. To find Δ_{12} we form the matrix

$$\begin{bmatrix} 1 & 0 & 1 & 1 \\ 0 & 1 & 1 & -1 \end{bmatrix}$$

Row reduce until the left hand block is I_2, which is already the case. On the other hand, to find Δ_{21} we form the matrix

$$\begin{bmatrix} 1 & 1 & 1 & 0 \\ 1 & -1 & 0 & 1 \end{bmatrix}$$

and row reduce, yielding the matrix

$$\begin{bmatrix} 1 & 0 & 1/2 & 1/2 \\ 0 & 1 & 1/2 & -1/2 \end{bmatrix}$$

A quick check verifies that indeed

$$\begin{bmatrix} 1 & 1 \\ 1 & -1 \end{bmatrix} \cdot \begin{bmatrix} 1/2 & 1/2 \\ 1/2 & -1/2 \end{bmatrix} = \begin{bmatrix} 1 & 0 \\ 0 & 1 \end{bmatrix}$$

Exercise 1.2.14 Show that the change of basis algorithm allows us to find the inverse of an $n \times n$ matrix A as follows: construct the $n \times 2n$ matrix $[A|I_n]$ and apply elementary row operations. If this results in a matrix $[I_n|B]$, then $B = A^{-1}$, if not, then A is not invertible. ◇

1.3 Diagonalization, Webpage Ranking, Data and Covariance

In this section, we develop the tools to analyze the smoking situation in Smallville. This will enable us to answer the questions posed earlier:

- What happens after n years?
- Is there an equilibrium state?

The key idea is that a matrix represents a linear transformation with respect to a basis, so by choosing a different basis, we may get a "better" representation. Our goal is to take a square matrix A and compute

$$\lim_{t \to \infty} A^t$$

So if "Tout est pour le mieux dans le meilleur des mondes possibles", perhaps we can find a basis where A is diagonal. Although Candide is doomed to disappointment, we are not! In many situations, we get lucky, and A can be diagonalized. To tackle this, we switch from French to German.

1.3.1 Eigenvalues and Eigenvectors

Suppose

$$T : V \to V$$

is a linear transformation; for concreteness let $V = \mathbb{R}^n$. If there exists a set of vectors $B = \{\mathbf{v}_1, \ldots, \mathbf{v}_n\}$ with

$$T(\mathbf{v}_i) = \lambda_i \mathbf{v}_i, \text{ with } \lambda_i \in \mathbb{R},$$

such that B is a basis for V, then the matrix M_{BB} representing T with respect to B (which is a basis for both source and target of T) is of the form

$$M_{BB} = \begin{bmatrix} \lambda_1 & 0 & 0 & 0 & 0 \\ 0 & \lambda_2 & 0 & 0 & 0 \\ 0 & 0 & \ddots & 0 & 0 \\ 0 & 0 & 0 & \ddots & 0 \\ 0 & 0 & 0 & 0 & \lambda_n \end{bmatrix}$$

This is exactly what happened in Example 1.1.8, and our next task is to determine how to find such a lucky basis. Given a matrix A representing T, we want to find vectors $\mathbf{v}$ and scalars λ satisfying

$$A\mathbf{v} = \lambda\mathbf{v} \text{ or equivalently } (A - \lambda \cdot I_n) \cdot \mathbf{v} = 0$$

The *kernel* of a matrix M is the set of $\mathbf{v}$ such that $M \cdot \mathbf{v} = 0$, so we need the kernel of $(A - \lambda \cdot I_n)$. Since the determinant of a square matrix is zero exactly when the matrix has a nonzero kernel, this means we need to solve for λ in the equation $\det(A - \lambda \cdot I_n) = 0$. The corresponding solutions are the *eigenvalues* of the matrix A.

Example 1.3.1 Let A be the matrix from Example 1.1.8:

$$\det \begin{bmatrix} 7 - \lambda & 2 \\ 3 & 8 - \lambda \end{bmatrix} = (7 - \lambda)(8 - \lambda) - 6 = \lambda^2 - 15\lambda + 50 = (\lambda - 5)(\lambda - 10).$$

So the eigenvalues of A are 5 and 10. These are exactly the values that appear on the diagonal of the matrix D; as we shall see, this is no accident.

For a given eigenvalue λ, we must find some nontrivial vector $\mathbf{v}$ which solves $(A - \lambda \cdot I)\mathbf{v} = 0$; these vectors are the *eigenvectors* of A. For this, we go back to solving systems of linear equations

Example 1.3.2 Staying in Smallville, we plug in our eigenvalues $\lambda \in \{5, 10\}$. First we solve for $\lambda = 5$:

$$\begin{bmatrix} 7-5 & 2 \\ 3 & 8-5 \end{bmatrix} \cdot \mathbf{v} = 0$$

which row reduces to the system

$$\begin{bmatrix} 1 & 1 \\ 0 & 0 \end{bmatrix} \cdot \mathbf{v} = 0,$$

which has as solution any nonzero multiple of $[1, -1]^T$. A similar computation for $\lambda = 10$ yields the eigenvector $[2, 3]^T$.

We've mentioned that the eigenvalues appear on the diagonal of the matrix D. Go back and take a look at Example 1.1.8 and see if you can spot how the eigenvectors come into play. If you get stuck, no worries: we tackle this next.

1.3.2 Diagonalization

We now revisit the change of basis construction. Let $T : V \rightarrow V$ be a linear transformation, and suppose we have two bases B_1 and B_2 for V. What is the relation between

$$M_{B_1 B_1} \text{ and } M_{B_2 B_2}?$$

The matrix $M_{B_1 B_1}$ takes as input a vector $\mathbf{v}_{B_1}$ written in terms of the B_1 basis, applies the operation T, and outputs the result in terms of the B_1 basis. We described change of basis as analogous to translation. To continue with this analogy, suppose T represents a recipe, B_1 is French and B_2 is German. Chef Pierre is French, and diner Hans is German. So Hans places his order $\mathbf{v}_{B_2}$ in German. Chef Pierre is temperamental—the waiter dare not pass on an order in German—so the order must be translated to French:

$$\mathbf{v}_{B_1} = \Delta_{12} \mathbf{v}_{B_2}.$$

This is relayed to Pierre, who produces

$$(M_{B_1 B_1}) \cdot \Delta_{12} \mathbf{v}_{B_2}.$$

Alas, Hans also has a short fuse, so the beleaguered waiter needs to present the dish to Hans with a description in German, resulting in

$$(\Delta_{21}) \cdot (M_{B_1 B_1}) \cdot (\Delta_{12}) \mathbf{v}_{B_2}.$$

Et voila! Comity in the restaurant. The reader unhappy with culinary analogies (or levity) should ignore the verbiage above, but keep the formulas.

Definition 1.3.3 Matrices A and B are *similar* if there is a C so $CAC^{-1} = B$.

Example 1.3.4 Let $B_1 = \{[1,0],[0,1]\}$ be the standard basis for $\mathbb{R}^2$, and $B_2 = \{[1,-1],[2,3]\}$ be the basis of eigenvectors we computed in Example 1.3.2. Using the algorithm of Exercise 1.2.14, we have that

$$\begin{bmatrix} 1 & 2 & 1 & 0 \\ -1 & 3 & 0 & 1 \end{bmatrix} \text{ row reduces to } \begin{bmatrix} 1 & 0 & 3/5 & -2/5 \\ 0 & 1 & 1/5 & 1/5 \end{bmatrix}$$

and a check shows that

$$\begin{bmatrix} 3/5 & -2/5 \\ 1/5 & 1/5 \end{bmatrix} \cdot \begin{bmatrix} 1 & 2 \\ -1 & 3 \end{bmatrix} = I_2.$$

These are exactly the matrices B and B^{-1} which appear in Example 1.1.8. Let's check our computation:

$$\begin{bmatrix} 3/5 & -2/5 \\ 1/5 & 1/5 \end{bmatrix} \cdot \begin{bmatrix} 7 & 2 \\ 3 & 8 \end{bmatrix} \cdot \begin{bmatrix} 1 & 2 \\ -1 & 3 \end{bmatrix} = \begin{bmatrix} 5 & 0 \\ 0 & 10 \end{bmatrix}$$

So we find

$$\lim_{t\to\infty} \begin{bmatrix} .7 & .2 \\ .3 & .8 \end{bmatrix}^t = \begin{bmatrix} 1 & 2 \\ -1 & 3 \end{bmatrix} \cdot \left(\lim_{t\to\infty} \begin{bmatrix} 1/2 & 0 \\ 0 & 1 \end{bmatrix}^t \right) \cdot \begin{bmatrix} 3/5 & -2/5 \\ 1/5 & 1/5 \end{bmatrix}$$

$$= \begin{bmatrix} 1 & 2 \\ -1 & 3 \end{bmatrix} \cdot \begin{bmatrix} 0 & 0 \\ 0 & 1 \end{bmatrix} \cdot \begin{bmatrix} 3/5 & -2/5 \\ 1/5 & 1/5 \end{bmatrix} = \begin{bmatrix} .4 & .4 \\ .6 & .6 \end{bmatrix}$$

Multiplying the last matrix by our start state vector $[n(0), s(0)]^T = [2000, 8000]^T$, we find the equilibrium state is $[4000, 6000]^T$. It is interesting that although we began with far more smokers than nonsmokers, and even though every year the percentage of nonsmokers who began smoking was larger than the percentage of smokers who quit, nevertheless in the equilibrium state we have more nonsmokers than in the initial state.

Exercise 1.3.5 Show that the matrix

$$\begin{bmatrix} \cos(\theta) & -\sin(\theta) \\ \sin(\theta) & \cos(\theta) \end{bmatrix}$$

which rotates a vector in $\mathbb{R}^2$ counterclockwise by θ degrees has real eigenvalues only when $\theta = 0$ or $\theta = \pi$. ◇

Exercise 1.3.6 Even if we work over an algebraically closed field, not all matrices are diagonalizable. Show that the matrix

$$\begin{bmatrix} 1 & 0 \\ 1 & 1 \end{bmatrix}$$

cannot be diagonalized. ◇

1.3.3 Ranking Using Diagonalization

Diagonalization is the key tool in many web search engines. The first task is to determine the right structure to represent the web; we will use a weighted, directed graph. Vertices of the graph correspond to websites, and edges correspond to links. If website A has a link to website B, this is represented by a directed edge from vertex A to vertex B; if website A has l links to other pages, each directed edge is assigned weight $\frac{1}{l}$. The idea is that a browser viewing website A has (in the absence of other information) an equal chance of choosing to click on any of the l links. From the data of a weighted, directed graph on vertices $\{v_1, \ldots, v_n\}$ we construct an $n \times n$ matrix T. Let l_j be the number of links at vertex v_j. Then

$$T_{ij} = \begin{cases} \frac{1}{l_j} & \text{if vertex } j \text{ has a link to vertex } i \\ 0 & \text{if vertex } j \text{ has no link to vertex } i. \end{cases}$$

Example 1.3.7 Consider the graph Γ:

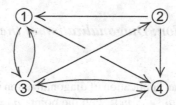

Using Γ, we construct the matrix

$$T = \begin{bmatrix} 0 & 1/2 & 1/3 & 0 \\ 0 & 0 & 1/3 & 0 \\ 1/2 & 0 & 0 & 1 \\ 1/2 & 1/2 & 1/3 & 0 \end{bmatrix}.$$

The matrix T is *column stochastic*: all column sums are one, which reflects the fact that the total probability at each vertex is one. There is a subtle point which makes this different than the Smallville smoking situation: a user may not choose to click on a link on the current page, but type in a URL. The solution is to add a second matrix R that represents the possibility of a *random jump*. Assuming that the user is equally likely to choose any of the n websites possible, this means that R is an $n \times n$ matrix with all entries $1/n$. Putting everything together, we have

$$G = (1 - p) \cdot T + p \cdot R, \text{ where } p \text{ is the probability of a random jump.}$$

The matrix G is called the *Google matrix*; experimental evidence indicates that p is close to .15. If a user is equally likely to start at any of the n vertices representing websites, the initial input vector to G is $\mathbf{v}(0) = [1/n, \ldots, 1/n]^T$. As in our previous vignette, we want to find

$$\mathbf{v}_\infty = \left(\lim_{t \to \infty} G^t \right) \cdot \mathbf{v}(0)$$

Theorem 1.3.8 [*Perron–Frobenius*] *If M is a column stochastic matrix with all entries positive, then 1 is an eigenvalue for M, and the corresponding eigenvector $\mathbf{v}_\infty$ has all positive entries, which sum to one. The limit above converges to $\mathbf{v}_\infty$.*

Therefore, one way of ranking webpages reduces to solving

$$G \cdot \mathbf{v} = I_n \cdot \mathbf{v}$$

The problem is that n is in the billions. In practice, rather than solving for $\mathbf{v}$ it is easier to get a rough approximation of $\mathbf{v}_\infty$ from $G^m \cdot \mathbf{v}(0)$ for m not too big.

1.3.4 Data Application: Diagonalization of the Covariance Matrix

In this section, we discuss an application of diagonalization in statistics. Suppose we have a data sample $X = \{p_1, \ldots, p_k\}$, with the points $p_i = (p_{i1}, \ldots, p_{im}) \in \mathbb{R}^m$. How do we visualize the spread of the data? Is it concentrated in certain directions?

Do large subsets cluster? Linear algebra provides one way to attack the problem. First, we need to define the covariance matrix: let $\mu_j(X)$ denote the mean of the jth coordinate of the points of X, and form the matrix

$$N_{ij} = p_{ij} - \mu_j(X)$$

The matrix N represents the original data, but with the points translated with respect to the mean in each coordinate.

Definition 1.3.9 The covariance matrix of X is $N^T \cdot N$.

Example 1.3.10 Consider the dataset

$$X = \{(1, 1), (2, 2), (2, 3), (3, 2), (3, 3), (4, 4)\}$$

Since $\sum_i p_{i1} = 15 = \sum_i p_{i2}$, we have $\mu_1(X) = 2.5 = \mu_2(X)$, so

$$N = \frac{1}{2} \begin{bmatrix} -3 & -3 \\ -1 & -1 \\ -1 & 1 \\ 1 & -1 \\ 1 & 1 \\ 3 & 3 \end{bmatrix}$$

The covariance matrix is therefore

$$N^T \cdot N = \frac{1}{4} \begin{bmatrix} -3 & -1 & -1 & 1 & 1 & 3 \\ -3 & -1 & 1 & -1 & 1 & 3 \end{bmatrix} \cdot \begin{bmatrix} -3 & -3 \\ -1 & -1 \\ -1 & 1 \\ 1 & -1 \\ 1 & 1 \\ 3 & 3 \end{bmatrix} = \begin{bmatrix} 11/2 & 9/2 \\ 9/2 & 11/2 \end{bmatrix}$$

To find the eigenvalues and eigenvectors for the covariance matrix, we compute:

$$\det \begin{bmatrix} 11/2 - \lambda & 9/2 \\ 9/2 & 11/2 - \lambda \end{bmatrix} = \lambda^2 - 11\lambda + 10 = (\lambda - 1)(\lambda - 10).$$

Exercise 1.3.11 Show if $\lambda = 1$, $\mathbf{v} = [1, -1]^T$ and that if $\lambda = 10$, $\mathbf{v} = [1, 1]^T$. ◇

Theorem 1.3.12 *Order the eigenvectors* $\{\lambda_1 \leq \lambda_2 \leq \ldots \leq \lambda_m\}$. *The data varies in proportion to the eigenvalues, in the direction of the associated eigenvector.*

The punchline is that the biggest eigenvalue corresponds to the biggest variance of the data, which "spreads out" in the direction of the corresponding eigenvector. Example 1.3.10 illustrates this nicely: the eigenvalue 10 corresponds to the direction $[1, 1]$, where the data spreads out the most, and the eigenvalue 1 corresponds to the direction $[1, -1]$, which is the second largest direction of spread. So for two dimensional data, the ellipse which best approximates the data is determined by the eigenvectors and eigenvalues of the covariance matrix; we compute the best fit ellipse in §A.1 using the statistics package R. Theorem 1.3.12 is the starting point of *principal component analysis*, which is a staple of applied mathematics.

1.4 Orthogonality, Least Squares Fitting, Singular Value Decomposition

The concept of orthogonality plays a key role in data science—generally it is not possible to perfectly fit data to reality, so the focus is on approximations. We want a good approximation, which will entail minimizing the distance between our approximation and the exact answer. The squared distance between two points is a quadratic function, so taking derivatives to minimize distance results in a system of linear equations. Vector calculus teaches us that minimization problems often involve projection onto a subspace. We now examine two fundamental tools in data analysis: least squares data fitting, and singular value decomposition. We start with a warm-up exercise on the law of cosines: for a triangle with side lengths A, B, C and opposite angles a, b, c, the law of cosines is

$$C^2 = A^2 + B^2 - 2AB\cos(c)$$

Exercise 1.4.1 Justify the assertion made in Definition 1.1.4 that the dot product is zero when vectors are orthogonal, as follows. Let the roles of A, B, C in the law of cosines be played by the lengths of vectors $\mathbf{v}, \mathbf{w}, \mathbf{w} - \mathbf{v}$, with $\mathbf{v}$ and $\mathbf{w}$ both emanating from the origin and c the angle between them. Apply the law of cosines to show that

$$\mathbf{v} \cdot \mathbf{w} = |\mathbf{v}| \cdot |\mathbf{w}| \cdot \cos(c).$$

Since $cos(c) = 0$ only when $c \in \{\frac{\pi}{2}, \frac{3\pi}{2}\}$, the result follows.   ◇

1.4.1 Least Squares

Let $X = \{p_1, \ldots, p_n\}$ be a set of data points in $\mathbb{R}^m$, and suppose we want to fit a curve (if $m = 2$) or a surface (if $m = 3$) or some other geometric structure to the data. If we allow too much freedom for our geometric object, the result is usually not a good approximation: for example, if $m = 2$ then since the space of polynomials of degree at most k has dimension $\binom{k+2}{2}$, there is a polynomial $f(x, y)$ such that $f(p_i) = 0$ for all $p_i \in X$ as soon as $\binom{k+2}{2} > n$. However, this polynomial often has lots of oscillation away from the points of X. So we need to make the question more precise, and develop criteria to evaluate what makes a fit "good".

Example 1.4.2 Consider a data set consisting of points in $\mathbb{R}^2$

$$X = \{(1, 6), (2, 5), (3, 7), (4, 10)\}$$

What *line* best approximates the data? Since a line will be given by the equation $y = ax + b$, the total error in the approximation will be

$$\text{error} = \sqrt{\sum_{(x_i, y_i) \in X} (y_i - (ax_i + b))^2}$$

For this example, the numbers are

x	y	$ax + b$	error
1	6	$a + b$	$(6 - (a + b))^2$
2	5	$2a + b$	$(5 - (2a + b))^2$
3	7	$3a + b$	$(7 - (3a + b))^2$
4	10	$4a + b$	$(10 - (4a + b))^2$

Minimizing $\sqrt{f}$ is equivalent to minimizing f, so we need to minimize

$$\begin{aligned} f(a, b) &= (6 - (a + b))^2 + (5 - (2a + b))^2 + (7 - (3a + b))^2 + (10 - (4a + b))^2 \\ &= 30a^2 + 20ab + 4b^2 - 154a - 56b + 210 \end{aligned}$$

which has partials

$$\begin{aligned}
\partial f/\partial a &= 60a + 20b - 154 \\
\partial f/\partial b &= 8b + 20a - 56
\end{aligned}$$

which vanishes when $b = 3.5$, $a = 1.4$. However, there is another way to look at this problem: we want to find the *best* solution to the system

$$\begin{bmatrix} 1 & 1 \\ 1 & 2 \\ 1 & 3 \\ 1 & 4 \end{bmatrix} \cdot \begin{bmatrix} b \\ a \end{bmatrix} = \begin{bmatrix} 6 \\ 5 \\ 7 \\ 10 \end{bmatrix}$$

This is equivalent to asking for the vector in the span of the columns of the left hand matrix which is closest to $[6, 5, 7, 10]^T$.

To understand how the approach to minimizing using partials relates to finding the closest vector in a subspace, we need to look more deeply at orthogonality.

1.4.2   Subspaces and Orthogonality

Definition 1.4.3 A subspace V' of a vector space V is a subset of V, which is itself a vector space. Subspaces W and W' of V are orthogonal if

$$\mathbf{w} \cdot \mathbf{w}' = 0 \text{ for all } \mathbf{w} \in W, \mathbf{w}' \in W'$$

A set of vectors $\{\mathbf{v}_1, \ldots \mathbf{v}_n\}$ is *orthonormal* if $\mathbf{v}_i \cdot \mathbf{v}_j = \delta_{ij}$ (1 if $i = j$, 0 otherwise). A matrix A is orthonormal when the column vectors of A are orthonormal.

Example 1.4.4 Examples of subspaces. Let A be an $m \times n$ matrix.

(a) The row space $R(A)$ of A is the span of the row vectors of A.
(b) The column space $C(A)$ of A is the span of the column vectors of A.
(c) The null space $N(A)$ of A is the set of solutions to the system $A \cdot \mathbf{x} = \mathbf{0}$.
(d) If W is a subspace of V, then $W^{\perp} = \{\mathbf{v} | \mathbf{v} \cdot \mathbf{w} = 0 \text{ for all } \mathbf{w} \in W\}$.

Notice that $\mathbf{v} \in N(A)$ iff $\mathbf{v} \cdot \mathbf{w} = 0$ for every row vector of A.

Exercise 1.4.5 Properties of important subspaces for A as above.

(a) Prove $N(A) = R(A)^{\perp}$, and $N(A^T) = C(A)^{\perp}$.
(b) Prove that any vector $\mathbf{v} \in V$ has a unique decomposition as $\mathbf{v} = \mathbf{w}' + \mathbf{w}''$ with $\mathbf{w}' \in W$ and $\mathbf{w}'' \in W^{\perp}$.
(c) Prove that for a vector $\mathbf{v} \in V$, the vector $\mathbf{w} \in W$ which minimizes $|\mathbf{v} - \mathbf{w}|$ is the vector $\mathbf{w}'$ above.

The *rank* of A is $\dim C(A) = \dim R(A)$. Show $n = \dim N(A) + \dim R(A)$.   ◇

Returning to the task at hand, the goal is to find the $\mathbf{x}$ that minimizes $|A\mathbf{x} - \mathbf{b}|$, which is equivalent to finding the vector in $C(A)$ closest to $\mathbf{b}$. By Exercise 1.4.5, we can write $\mathbf{b}$ uniquely as $\mathbf{b}' + \mathbf{b}''$ with $\mathbf{b}' \in C(A)$ and $\mathbf{b}'' \in C(A)^{\perp}$. Continuing to reap the benefits of Exercise 1.4.5, we have

$$\mathbf{b} = \mathbf{b}' + \mathbf{b}'' \Rightarrow A^T\mathbf{b} = A^T\mathbf{b}' + A^T\mathbf{b}'' = A^TA \cdot \mathbf{y} + 0,$$

where the last equality follows because $\mathbf{b}' \in C(A) \Rightarrow \mathbf{b}' = A \cdot \mathbf{y}$, and since $\mathbf{b}'' \in C(A)^{\perp} \Rightarrow A^T\mathbf{b}'' = 0$. Thus

$$A^T\mathbf{b} = A^TA \cdot \mathbf{y} \text{ and therefore } \mathbf{y} = (A^TA)^{-1}A^T\mathbf{b}$$

solves the problem, as long as A^TA is invertible. In Exercise 1.1.7 we showed that A^TA is symmetric; it turns out that symmetric matrices are exactly those which have the property that they can be diagonalized by an orthonormal change of basis matrix B, in which case we have $B^T = B^{-1}$. This fact is called the *spectral theorem*. One direction is easy: suppose A is a matrix having such a change of basis. Then $(B^{-1})^T = (B^T)^T = B$, so

$$BAB^{-1} = D = D^T = (BAB^{-1})^T = BA^TB^{-1} \Rightarrow A = A^T$$

Exercise 1.4.6 Show that a real symmetric matrix has only real eigenvalues, and this need not hold for an arbitrary real matrix. Now use induction to prove that if A is symmetric, it admits an orthonormal change of basis matrix as above. $\diamond$

Example 1.4.7 Let A denote the matrix on the left in the last displayed equation in Example 1.4.2, and let $\mathbf{b} = [6, 5, 7, 10]^T$. Then

$$A^TA = \begin{bmatrix} 1 & 1 & 1 & 1 \\ 1 & 2 & 3 & 4 \end{bmatrix} \cdot \begin{bmatrix} 1 & 1 \\ 1 & 2 \\ 1 & 3 \\ 1 & 4 \end{bmatrix} = \begin{bmatrix} 4 & 10 \\ 10 & 30 \end{bmatrix}$$

so

$$(A^TA)^{-1} = \begin{bmatrix} 3/2 & -1/2 \\ -1/2 & 1/5 \end{bmatrix}$$

Continuing with the computation, we have

$$A \cdot (A^TA)^{-1} \cdot A^T = \frac{1}{10} \begin{bmatrix} 7 & 4 & 1 & -2 \\ 4 & 3 & 2 & 1 \\ 1 & 2 & 3 & 4 \\ -2 & 1 & 4 & 7 \end{bmatrix}$$

Putting everything together, we see that indeed

$$A \cdot (A^T A)^{-1} \cdot A^T \cdot \mathbf{b} = \begin{bmatrix} 4.9 \\ 6.3 \\ 7.7 \\ 9.1 \end{bmatrix} = \begin{bmatrix} 1 & 1 \\ 1 & 2 \\ 1 & 3 \\ 1 & 4 \end{bmatrix} \cdot \begin{bmatrix} 3.5 \\ 1.4 \end{bmatrix}$$

where [3.5, 1.4] is the solution we obtained using partials.

Exercise 1.4.8 What happens when we fit a degree two equation $y = ax^2 + bx + c$
to the data of Example 1.4.2? The corresponding matrix equation is given by

$$\begin{bmatrix} 1 & 1 & 1 \\ 1 & 2 & 4 \\ 1 & 3 & 9 \\ 1 & 4 & 16 \end{bmatrix} \cdot \begin{bmatrix} c \\ b \\ a \end{bmatrix} = \begin{bmatrix} 6 \\ 5 \\ 7 \\ 10 \end{bmatrix}$$

Carry out the analysis conducted in Example 1.4.7 for this case, and show that the
solution obtained by solving for $\mathbf{y}$ and computing $\mathbf{b}' = A\mathbf{y}$ agrees with the solution
obtained by using partials. $\diamond$

1.4.3   Singular Value Decomposition

Diagonalization can provide elegant and computationally efficient solutions to
questions like that posed in Example 1.1.1. One drawback is that we are constrained
to square matrices, whereas the problems encountered in the real world often have
the target different from the source, so that the matrix or linear transformation in
question is not square. Singular Value Decomposition (SVD) is *"diagonalization
for non-square matrices"*.

Theorem 1.4.9 *Let M be an $m \times n$ matrix of rank r. Then there exist matrices U of
size $m \times m$ and V of size $n \times n$ with orthonormal columns, and Σ an $m \times n$ diagonal
matrix with nonzero entries $\Sigma_{ii} = \{\sigma_1, \ldots, \sigma_r\}$, such that*

$$M = U \Sigma V^T.$$

Proof The matrix $M^T M$ is symmetric, so by Exercise 1.4.6 there is an orthonormal
change of basis that diagonalizes $M^T M$. Let $M^T M \cdot \mathbf{v}_j = \lambda_j \mathbf{v}_j$, and note that

$$\mathbf{v}_i^T M^T M \mathbf{v}_j = \lambda_j \mathbf{v}_i^T \mathbf{v}_j = \lambda_j \delta_{ij}$$

Define $\sigma_i = \sqrt{\lambda_i}$, and $\mathbf{q}_i = \frac{1}{\sigma_i} M \mathbf{v}_i$. Then

$$\mathbf{q}_i^T \mathbf{q}_j = \delta_{ij} \text{ for } j \in \{1, \ldots, r\}.$$

Extend the $\mathbf{q}_i$ to a basis for $\mathbb{R}^m$, and let U be the matrix whose columns are the $\mathbf{q}_i$, and V the matrix whose columns are the $\mathbf{v}_i$. Then

$$(U^T M V)_{ij} = \mathbf{q}_i^T (MV)_{col_j} = \mathbf{q}_i^T M \mathbf{v}_j = \sigma_j \mathbf{q}_i^T \mathbf{q}_j = \sigma_j \delta_{ij} \text{ if } j \leq r, \text{ or } 0 \text{ if } j > r.$$

Hence $U^T M V = \Sigma$; the result follows using that $U^T = U^{-1}$ and $V^T = V^{-1}$. □

Example 1.4.10 We compute the Singular Value Decomposition for

$$M = \begin{bmatrix} 1 & 2 \\ 0 & 1 \\ 2 & 0 \end{bmatrix}$$

First, we compute

$$M^T M = \begin{bmatrix} 5 & 2 \\ 2 & 5 \end{bmatrix}$$

We find (do it!) that the eigenvalues are $\{7, 3\}$ with corresponding eigenvectors $\{[1, 1]^T, [1, -1]^T\}$. The eigenvectors are the columns of V^T; they both have length $\sqrt{2}$ so to make them orthonormal we scale by $\frac{1}{\sqrt{2}}$, and we have

$$\Sigma = \begin{bmatrix} \sqrt{7} & 0 \\ 0 & \sqrt{3} \\ 0 & 0 \end{bmatrix} \text{ and } V = \frac{1}{\sqrt{2}} \begin{bmatrix} 1 & 1 \\ 1 & -1 \end{bmatrix}$$

It remains to calculate U, which we do using the recipe $\mathbf{u}_i = \frac{1}{\sigma_i} \cdot M \mathbf{v}_i$. This gives us

$$\mathbf{u}_1 = \frac{1}{\sqrt{14}} \cdot \begin{bmatrix} 3 \\ 1 \\ 2 \end{bmatrix} \text{ and } \mathbf{u}_2 = \frac{1}{\sqrt{6}} \cdot \begin{bmatrix} -1 \\ -1 \\ 2 \end{bmatrix}$$

We need to extend this to an orthonormal basis for $\mathbb{R}^3$, so we need to calculate a basis for $N(M^T) = C(M)^\perp$, which we find consists of $\mathbf{u}_3 = \frac{1}{\sqrt{21}}[-2, 4, 1]^T$, hence

$$U = \begin{bmatrix} \frac{3}{\sqrt{14}} & \frac{-1}{\sqrt{6}} & \frac{-2}{\sqrt{21}} \\ \frac{1}{\sqrt{14}} & \frac{-1}{\sqrt{6}} & \frac{4}{\sqrt{21}} \\ \frac{2}{\sqrt{14}} & \frac{2}{\sqrt{6}} & \frac{1}{\sqrt{21}} \end{bmatrix}$$

Exercise 1.4.11 Check that indeed $M = U\Sigma V^T$. What is the rank one matrix which best approximates M? ◇

The utility of SVD stems from the fact that it allows us to represent the matrix M as a sum of simpler matrices, in particular, matrices of rank one.

Example 1.4.12

$$M = \mathbf{u}_1\sigma_1\mathbf{v}_1^T + \mathbf{u}_2\sigma_2\mathbf{v}_2^T + \cdots + \mathbf{u}_r\sigma_r\mathbf{v}_r^T$$

which means we can decompose M as a sum of rank one matrices; the bigger the value of σ_i, the larger the contribution to M. For example, a greyscale image is comprised of $m \times n$ pixels, and by doing an SVD decomposition, we can get a good approximation to the image which takes up very little storage by keeping only the highest weight terms.

Exercise 1.4.13 Least squares approximation is also an instance of SVD. Recall that in least squares, we are trying to find the best approximate solution to a system $M\mathbf{x} = \mathbf{b}$. So our goal is to minimize $|M\mathbf{x} - \mathbf{b}|$. We have

$$
\begin{aligned}
|M\mathbf{x} - \mathbf{b}| &= |U\Sigma V^T\mathbf{x} - \mathbf{b}| \\
&= |\Sigma V^T\mathbf{x} - U^T\mathbf{b}| \\
&= |\Sigma \mathbf{y} - U^T\mathbf{b}|
\end{aligned}
$$

Show this is minimized by

$$\mathbf{y} = V^T\mathbf{x} = \frac{1}{\Sigma}U^T\mathbf{b},$$

where we write $\frac{1}{\Sigma}$ to denote a matrix with $\frac{1}{\sigma_i}$ on the diagonal. ◇

Chapter 2
Basics of Algebra: Groups, Rings, Modules

The previous chapter covered linear algebra, and in this chapter we move on to more advanced topics in abstract algebra, starting with the concepts of group and ring. In Chap. 1, we defined a vector space over a field $\mathbb{K}$ without giving a formal definition for a field; this is rectified in §1. A field is a specific type of ring, so we will define a field in the context of a more general object. The reader has already encountered many rings besides fields: the integers are a ring, as are polynomials in one or more variables, and square matrices.

When we move into the realm of topology, we'll encounter more exotic rings, such as differential forms. Rings bring a greater level of complexity into the picture, and with that, the ability to build structures and analyze objects in finer detail. For instance, Example 2.2.8 gives a first glimpse of the objects appearing in persistent homology, which is the centerpiece of Chap. 7. In this chapter, we'll cover

- Groups, Rings and Homomorphisms.
- Modules and Operations on Modules.
- Localization of Rings and Modules.
- Noetherian Rings, Hilbert Basis Theorem, Variety of an Ideal.

2.1 Groups, Rings and Homomorphisms

2.1.1 Groups

Let G be a set of elements endowed with a binary operation sending $G \times G \longrightarrow G$ via $(a, b) \mapsto a \cdot b$; in particular G is closed under the operation. The set G is a group if the following three properties hold:

- $\cdot$ is associative: $(a \cdot b) \cdot c = a \cdot (b \cdot c)$.
- G possesses an identity element e such that $\forall\ g \in G,\ g \cdot e = e \cdot g = g$.
- Every $g \in G$ has an inverse g^{-1} such that $g \cdot g^{-1} = g^{-1} \cdot g = e$.

© The Author(s), under exclusive license to Springer Nature Switzerland AG 2022
H. Schenck, *Algebraic Foundations for Applied Topology and Data Analysis*,
Mathematics of Data 1, https://doi.org/10.1007/978-3-031-06664-1_2

The group operation is commutative if $a \cdot b = b \cdot a$; in this case the group is *abelian*. The prototypical example of an abelian group is the set of all integers, with addition serving as the group operation. For abelian groups it is common to write the group operation as $+$. In a similar spirit, $\mathbb{Z}/n\mathbb{Z}$ (often written $\mathbb{Z}/n$ for brevity) is an abelian group with group operation addition modulo n. If instead of addition in $\mathbb{Z}$ we attempt to use multiplication as the group operation, we run into a roadblock: zero clearly has no inverse.

For an example of a group which is not abelian, recall that matrix multiplication is not commutative. The set G of $n \times n$ *invertible* matrices with entries in $\mathbb{R}$ is a non-abelian group if $n \geq 2$, with group operation matrix multiplication.

Exercise 2.1.1 Determine the multiplication table for the group of 2×2 invertible matrices with entries in $\mathbb{Z}/2$. There are 16 2×2 matrices with $\mathbb{Z}/2$ entries, but any matrix of rank zero or rank one is not a member. You should find 6 elements. $\diamond$

Definition 2.1.2 A *subgroup* of a group G is a subset of G which is itself a group. A subgroup H of G is *normal* if $gHg^{-1} \subseteq H$ for all $g \in G$, where gHg^{-1} is the set of elements ghg^{-1} for $h \in H$. A *homomorphism* of groups is a map which preserves the group structure, so a map $G_1 \xrightarrow{f} G_2$ is a homomorphism if

$$f(g_1 \cdot g_2) = f(g_1) \cdot f(g_2) \text{ for all } g_i \in G_i.$$

The *kernel* of f is the set of $g \in G_1$ such that $f(g) = e$. If the kernel of f is $\{e\}$ then f is *injective* or *one-to-one*; if every $g \in G_2$ has $g = f(g')$ for some $g' \in G_1$ then f is *surjective* or *onto*, and if f is both injective and surjective, then f is an *isomorphism*.

Exercise 2.1.3 First, prove that the kernel of a homomorphism is a normal subgroup. Next, let

$$G/H = \text{ the set of equivalence classes, with } a \sim b \text{ iff } aH = bH \text{ iff } ab^{-1} \in H.$$

Prove that the condition $gHg^{-1} \subseteq H$ that defines a normal subgroup insures that the *quotient* G/H is itself a group, as long as H is normal. $\diamond$

Exercise 2.1.4 Label the vertices of an equilateral triangle as $\{1, 2, 3\}$, and consider the rigid motions, which are rotations by integer multiples of $\frac{2\pi}{3}$ and reflection about any line connecting a vertex to the midpoint of the opposite edge. Prove that this group has 6 elements, and is isomorphic to the group in Exercise 2.1.1, as well as to the group S_3 of permutations of three letters, with the group operation in S_3 given by composition: if $\sigma(1) = 2$ and $\tau(2) = 3$, then $\tau \cdot \sigma(1) = 3$. $\diamond$

Definition 2.1.5 There are a number of standard operations on groups:

- Direct Product and Direct Sum.
- Intersection.
- Sum (when Abelian).

The direct product of groups G_1 and G_2 consists of pairs (g_1, g_2) with $g_i \in G_i$, and group operation defined pointwise as below; we write the direct product as $G_1 \oplus G_2$.

$$(g_1, g_2) \cdot (g_1', g_2') = (g_1 g_1', g_2 g_2').$$

This extends to a finite number of G_i, and in this case is also known as the direct sum. To form intersection and sum, the groups G_1 and G_2 must be subgroups of a larger ambient group G, and are defined as below. For the sum we will write the group operation additively as $G_1 + G_2$; the groups we encounter are all Abelian.

$$
\begin{aligned}
G_1 \cap G_2 &= \{g \mid g \in G_1 \text{ and } g \in G_2\} \\
G_1 + G_2 &= \{g_1 + g_2 \text{ for some } g_i \in G_i\}
\end{aligned}
$$

Exercise 2.1.6 Show that the constructions in Definition 2.1.5 do yield groups. ◇

2.1.2 Rings

A *ring R* is an abelian group under addition (which henceforth will always be written as $+$, with additive identity written as 0), with an additional associative operation multiplication ($\cdot$) which is distributive with respect to addition. An additive subgroup $I \subseteq R$ such that $r \cdot i \in I$ for all $r \in R, i \in I$ is an *ideal*. In these notes, unless otherwise noted, rings will have

- a multiplicative identity, written as 1.
- commutative multiplication.

Example 2.1.7 Examples of rings not satisfying the above properties:

- As a subgroup of $\mathbb{Z}$, the even integers $2\mathbb{Z}$ satisfy the conditions above for the usual multiplication and addition, but have no multiplicative identity.
- The set of all 2×2 matrices over $\mathbb{R}$ is an abelian group under $+$, has a multiplicative identity, and satisfies associativity and distributivity. But multiplication is not commutative.

Definition 2.1.8 A *field* is a commutative ring with unit $1 \neq 0$, such that every nonzero element has a multiplicative inverse. A nonzero element a of a ring is a *zero divisor* if there is a nonzero element b with $a \cdot b = 0$. An *integral domain* (for brevity, *domain*) is a ring with no zero divisors.

Remark 2.1.9 General mathematical culture: a noncommutative ring such that every nonzero element has a left and right inverse is called a *division ring*. The most famous example is the ring of quaternions, discovered by Hamilton in 1843 and etched into the Brougham Bridge.

Example 2.1.10 Examples of rings.

(a) $\mathbb{Z}$, the integers, and $\mathbb{Z}/n\mathbb{Z}$, the integers mod n.
(b) $A[x_1, \ldots, x_n]$, the polynomials with coefficients in a ring A.
(c) $C^0(\mathbb{R})$, the continuous functions on $\mathbb{R}$.
(d) $\mathbb{K}$ a field.

Definition 2.1.11 If R and S are rings, a map $\phi : R \to S$ is a ring homomorphism if it respects the ring operations: for all $r, r' \in R$,

(a) $\phi(r \cdot r') = \phi(r) \cdot \phi(r')$.
(b) $\phi(r + r') = \phi(r) + \phi(r')$.
(c) $\phi(1) = 1$.

Example 2.1.12 There is no ring homomorphism from

$$\mathbb{Z}/2 \xrightarrow{\phi} \mathbb{Z}.$$

To see this, note that in $\mathbb{Z}/2$ the zero element is 2. Hence in $\mathbb{Z}$ we would have

$$0 = \phi(2) = \phi(1) + \phi(1) = 1 + 1 = 2$$

An important construction is that of the *quotient ring*:

Definition 2.1.13 Let $I \subseteq R$ be an ideal. Elements of the quotient ring R/I are equivalence classes, with $(r + [I]) \sim (r' + [I])$ if $r - r' \in I$.

$$
\begin{aligned}
(r + [I]) \cdot (r' + [I]) &= r \cdot r' + [I] \\
(r + [I]) + (r' + [I]) &= r + r' + [I]
\end{aligned}
$$

Note a contrast with the construction of a quotient group: for H a subgroup of G, G/H is itself a group only when H is a normal subgroup. There is not a similar constraint on a ring quotient, because the additive operation in a ring is commutative, so all additive subgroups of R (in particular, ideals) are abelian, hence are normal subgroups with respect to the additive structure.

Exercise 2.1.14 Prove the kernel of a ring homomorphism is an ideal.   ◇

2.2 Modules and Operations on Modules

In linear algebra, we can add two vectors together, or multiply a vector by an element of the field over which the vector space is defined. Module is to ring what vector space is to field. We saw above that a field is a special type of ring; a module over a field is a vector space. We generally work with commutative rings; if this is not the case (e.g. tensor and exterior algebras) we make note of the fact.

Definition 2.2.1 A *module* M over a ring R is an abelian group, together with an action of R on M which is R-linear: for $r_i \in R$, $m_i \in M$,

- $r_1(m_1 + m_2) = r_1 m_1 + r_1 m_2$,
- $(r_1 + r_2)m_1 = r_1 m_1 + r_2 m_1$,
- $r_1(r_2 m_1) = (r_1 r_2)m_1$,
- $1 \cdot (m) = m$.
- $0 \cdot (m) = 0$.

An R-module M is *finitely-generated* if there exist $S = \{m_1, \ldots, m_n\} \subseteq M$ such that any $m \in M$ can be written

$$m = \sum_{i=1}^{n} r_i m_i$$

for some $r_i \in R$. We write $\langle m_1, \ldots, m_n \rangle$ to denote that S is a set of generators.

Exercise 2.2.2 A subset $N \subseteq M$ of an R-module M is a *submodule* if N is itself an R-module; a submodule $I \subset R$ is called an *ideal*. Which sets below are ideals?

(a) $\{f \in C^0(\mathbb{R}) \mid f(1) = 0\}$.
(b) $\{f \in C^0(\mathbb{R}) \mid f(1) \neq 0\}$.
(c) $\{n \in \mathbb{Z} \mid n = 0 \bmod 3\}$.
(d) $\{n \in \mathbb{Z} \mid n \neq 0 \bmod 3\}$.

Example 2.2.3 Examples of modules over a ring R.

(a) Any ring is a module over itself.
(b) A quotient ring R/I is both an R-module and an R/I-module.
(c) An R-module M is *free* if it is free of relations among a minimal set of generators, so a finitely generated free R-module is isomorphic to R^n. An ideal with two or more generators is not free: if $I = \langle f, g \rangle$ then we have the trivial (or *Koszul*) relation $f \cdot g - g \cdot f = 0$ on I. For modules M and N, Definition 2.1.5 shows we can construct a new module via the direct sum $M \oplus N$. If $M, N \subseteq P$ we also have modules $M \cap N$ and $M + N$.

Definition 2.2.4 Let M_1 and M_2 be modules over R, $m_i \in M_i$, $r \in R$. A homomorphism of R-modules $\psi : M_1 \to M_2$ is a function ψ such that

(a) $\psi(m_1 + m_2) = \psi(m_1) + \psi(m_2)$.
(b) $\psi(r \cdot m_1) = r \cdot \psi(m_1)$.

Notice that when $R = \mathbb{K}$, these two conditions are exactly those for a linear transformation, again highlighting that *module* is to *ring* as *vector space* is to *field*.

Definition 2.2.5 Let

$$M_1 \xrightarrow{\phi} M_2$$

be a homomorphism of R-modules.

- The kernel $\ker(\phi)$ of ϕ consists of those $m_1 \in M_1$ such that $\phi(m_1) = 0$.
- The image $\mathrm{im}(\phi)$ of ϕ consists of those $m_2 \in M_2$ such that $m_2 = \phi(m_1)$ for some $m_1 \in M_1$.
- The cokernel $\mathrm{coker}(\phi)$ of ϕ consists of $M_2/\mathrm{im}(M_1)$.

Exercise 2.2.6 Prove that the kernel, image, and cokernel of a homomorphism of R-modules are all R-modules.   ◇

Definition 2.2.7 A sequence of R–modules and homomorphisms

$$\mathscr{C}: \quad \cdots \xrightarrow{\phi_{j+2}} M_{j+1} \xrightarrow{\phi_{j+1}} M_j \xrightarrow{\phi_j} M_{j-1} \xrightarrow{\phi_{j-1}} \cdots \qquad (2.2.1)$$

is a *complex* (or chain complex) if

$$\mathrm{im}(\phi_{j+1}) \subseteq \ker(\phi_j).$$

The sequence is *exact* at M_j if $\mathrm{im}(\phi_{j+1}) = \ker(\phi_j)$; a complex which is exact everywhere is called an *exact sequence*. The jth homology module of $\mathscr{C}$ is:

$$H_j(\mathscr{C}) = \ker(\phi_j)/\mathrm{im}(\phi_{j+1}).$$

An exact sequence of the form

$$\mathscr{C}: \quad 0 \xrightarrow{} A_2 \xrightarrow{d_2} A_1 \xrightarrow{d_1} A_0 \xrightarrow{} 0 \qquad (2.2.2)$$

is called a *short exact sequence*.

Exercise 2.2.8 This exercise foreshadows the key concept of algebraic topology, which is to use *algebra* to encode topological features of a space. Consider a trio of vector spaces over $\mathbb{K} = \mathbb{Z}/2$, of dimensions $\{1, 3, 3\}$. Let [012] be a basis for

A_2, $\{[01], [02], [12]\}$ a basis for A_1, and $\{[0], [1], [2]\}$ a basis for A_0. The labeling of basis elements is prompted by the pictures below; the map d_i represents the *boundaries* of elements of A_i.

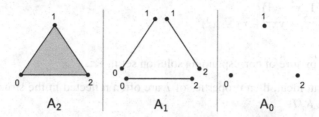

$$A_2 \qquad\qquad A_1 \qquad\qquad A_0$$

Show that if $d_2 = [1, 1, 1]^T$ and

$$d_1 = \begin{bmatrix} 1 & 1 & 0 \\ 1 & 0 & 1 \\ 0 & 1 & 1 \end{bmatrix}$$

then the sequence 2.2.2 is exact except at A_0. Now drop A_2 from 2.2.2, and show the complex below

$$0 \longrightarrow A_1 \xrightarrow{\ d_1\ } A_0 \longrightarrow 0$$

has $H_0 \simeq \mathbb{K} \simeq H_1$. We will return to this example in Chap. 4; roughly speaking the nonzero H_1 reflects the fact that a (hollow) triangle is topologically the same as S^1, which has a one dimensional "hole". In Exercise 4.2.7 of Chap. 4, you'll prove that $H_0 \simeq \mathbb{K}$ reflects the fact that S^1 is connected. $\Diamond$

2.2.1 Ideals

The most commonly encountered modules are ideals, which are submodules of the ambient ring R. An ideal $I \subseteq R$ is *proper* if $I \neq R$.

Definition 2.2.9 Types of proper ideals I

(a) I is *principal* if I can be generated by a single element.
(b) I is *prime* if $f \cdot g \in I$ implies either f or g is in I.
(c) I is *maximal* if there is no proper ideal J with $I \subsetneq J$.
(d) I is *primary* if $f \cdot g \in I \Rightarrow f$ or g^m is in I, for some $m \in \mathbb{N}$.
(e) I is *reducible* if there exist ideals J_1, J_2 such that $I = J_1 \cap J_2$, $I \subsetneq J_i$.
(f) I is *radical* if $f^m \in I$ ($m \in \mathbb{N} = \mathbb{Z}_{>0}$) implies $f \in I$.

Exercise 2.2.10 Which classes above do the ideals $I \subseteq \mathbb{R}[x, y]$ below belong to?

(a) $\langle xy \rangle$
(b) $\langle y - x^2, y - 1 \rangle$
(c) $\langle y, x^2 - 1, x^5 - 1 \rangle$
(d) $\langle y - x^2, y^2 - yx^2 + xy - x^3 \rangle$
(e) $\langle xy, x^2 \rangle$

Hint: draw a picture of corresponding solution set in $\mathbb{R}^2$. $\diamond$

If $I \subseteq R$ is an ideal, then properties of I are often reflected in the structure of the quotient ring R/I.

Theorem 2.2.11

$$R/I \text{ is a domain} \iff I \text{ is a prime ideal.}$$
$$R/I \text{ is a field} \iff I \text{ is a maximal ideal.}$$

Proof For the first part, R/I is a domain iff there are no zero divisors, hence $a \cdot b = 0$ implies $a = 0$ or $b = 0$. If $\tilde{a}$ and $\tilde{b}$ are representatives in R of a and b, then $a \cdot b = 0$ in R/I is equivalent to

$$\tilde{a}\tilde{b} \in I \iff \tilde{a} \in I \text{ or } \tilde{b} \in I,$$

which holds iff I is prime. For the second part, suppose R/I is a field, but I is not a maximal ideal, so there exists a proper ideal J satisfying

$$I \subset J \subseteq R.$$

Take $j \in J$. Since R/I is a field, there exists j' such that

$$(j' + [I]) \cdot (j + [I]) = 1 \text{ which implies } jj' + (j + j')[I] + [I] = 1.$$

But $(j + j')[I] + [I] = 0$ in R/I, so $jj' = 1$ in R/I, hence J is not a proper ideal, a contradiction. $\square$

Exercise 2.2.12 A *local ring* is a ring with a unique maximal ideal $\mathfrak{m}$. Prove that in a local ring, if $f \notin \mathfrak{m}$, then f is an invertible element (also called a *unit*). $\diamond$

Exercise 2.2.13 A ring is a *Principal Ideal Domain* (PID) if it is an integral domain, and every ideal is principal. In Chap. 6, we will use the Euclidean algorithm to show that $\mathbb{K}[x]$ is a PID. Find a generator for

$$\langle x^4 - 1, x^3 - 3x^2 + 3x - 1 \rangle.$$

Is $\mathbb{K}[x, y]$ a PID? $\diamond$

2.2.2 Tensor Product

From the viewpoint of the additive structure, a module is just an abelian group, hence the operations of intersection, sum, and direct sum that we defined for groups can be carried out for modules M_1 and M_2 over a ring R. When M_1 is a module over R_1 and M_2 is a module over R_2, a homomorphism

$$R_1 \xrightarrow{\phi} R_2$$

allows us to give M_2 the structure of an R_1-module via

$$r_1 \cdot m_2 = \phi(r_1) \cdot m_2.$$

We can also use ϕ to make M_1 into an R_2-module via *tensor product*.

Definition 2.2.14 Let M and N be R-modules, and let P be the free R-module generated by $\{(m, n) | m \in M, n \in N\}$. Let Q be the submodule of P generated by

$$(m_1 + m_2, n) - (m_1, n) - (m_2, n)$$
$$(m, n_1 + n_2) - (m, n_1) - (m, n_2)$$
$$(rm, n) - r(m, n)$$
$$(m, rn) - r(m, n).$$

The tensor product is the R-module:

$$M \otimes_R N = P/Q.$$

We write $m \otimes n$ to denote the class (m, n).

This seems like a strange construction, but with a bit of practice, tensor product constructions become very natural. The relations $(rm, n) \sim r(m, n)$ and $(m, rn) \sim r(m, n)$ show that tensor product is R-linear.

Example 2.2.15 For a vector space V over $\mathbb{K}$, the tensor algebra $T(V)$ is a noncommutative ring, constructed iteratively as follows. Let $V^i = V \otimes V \otimes \cdots \otimes V$ be the i-fold tensor product, with $V^0 = \mathbb{K}$. Then

$$T(V) = \bigoplus_i V^i$$

is the tensor algebra. The symmetric algebra $Sym(V)$ is obtained by quotienting $T(V)$ by the relation $v_i \otimes v_j - v_j \otimes v_i = 0$. When $V = \mathbb{K}^n$, $Sym(V)$ is isomorphic to the polynomial ring $\mathbb{K}[x_1, \ldots, x_n]$. The exterior algebra $\Lambda(V)$ is defined in similar fashion, except that we quotient by the relation $v_i \otimes v_j + v_j \otimes v_i = 0$.

Exercise 2.2.16 For $a, b \in \mathbb{Z}$, show that

$$(\mathbb{Z}/a\mathbb{Z}) \otimes_{\mathbb{Z}} (\mathbb{Z}/b\mathbb{Z}) \simeq \mathbb{Z}/GCD(a,b)\mathbb{Z}.$$

In particular, when a and b are relatively prime, the tensor product is zero.   ◇

If M, N, and T are R-modules, then a map

$$M \times N \xrightarrow{f} T$$

is *bilinear* if $f(rm_1 + m_2, n) = rf(m_1, n) + f(m_2, n)$, and similarly in the second coordinate. Tensor product converts R–bilinear maps into R–linear maps, and possesses a *universal mapping property*: given a bilinear map f, there is a unique R–linear map $M \otimes_R N \longrightarrow T$ making the following diagram commute:

$$
\begin{array}{ccc}
M \times N & \xrightarrow{f} & T \\
\downarrow & & \\
M \otimes_R N & &
\end{array}
$$

Exercise 2.2.17 Prove the universal mapping property of tensor product.   ◇

The motivation to define tensor product was to give the R_1-module M_1 the structure of a R_2-module. This operation is known as *extension of scalars*. The map

$$R_1 \xrightarrow{\phi} R_2$$

makes R_2 into an R_1-module via $r_1 \cdot r_2 = \phi(r_1) \cdot r_2$, so we can tensor M_1 and R_2 over R_1 to obtain

$$M_1 \otimes_{R_1} R_2$$

which is both an R_1-module and an R_2-module.
What is the effect tensoring a short exact sequence of R-modules

$$0 \longrightarrow A_1 \longrightarrow A_2 \longrightarrow A_3 \longrightarrow 0$$

with an R-module M? It turns out that exactness is preserved at all but the leftmost position.

Exercise 2.2.18 Show that if M is an R-module and $A_\bullet$ is a short exact sequence as above, then

$$A_1 \otimes_R M \longrightarrow A_2 \otimes_R M \longrightarrow A_3 \otimes_R M \longrightarrow 0.$$

is exact. Show that if $R = \mathbb{Z}$, then tensoring the exact sequence

$$0 \longrightarrow \mathbb{Z} \xrightarrow{\cdot 2} \mathbb{Z} \longrightarrow \mathbb{Z}/2 \longrightarrow 0$$

with $\mathbb{Z}/2$ does not preserve exactness in the leftmost position. $\diamond$

2.2.3 Hom

For a pair of R-modules M_1 and M_2, the set of all R-module homomorphisms from M_1 to M_2 is itself an R-module, denoted $\text{Hom}_R(M_1, M_2)$. To determine the R-module structure on $\text{Hom}_R(M_1, M_2)$, we examine how an element acts on $m_1 \in M_1$:

$$(\phi_1 + \phi_2)(m_1) = \phi_1(m_1) + \phi_2(m_1) \text{ and } (r \cdot \phi_1)(m_1) = r \cdot \phi_1(m_1).$$

Given

$$\psi \in \text{Hom}_R(M_1, M_2) \text{ and } \phi \in \text{Hom}_R(M_2, N),$$

we can compose them:

$$\phi \circ \psi \in \text{Hom}_R(M_1, N).$$

Put differently, we can apply $\text{Hom}_R(\bullet, N)$ to input $M_1 \xrightarrow{\phi} M_2$, yielding output

$$\text{Hom}_R(M_2, N) \longrightarrow \text{Hom}_R(M_1, N) \text{ via } \psi \mapsto \phi \circ \psi.$$

When we applied $\bullet \otimes_R M$ to a short exact sequence, we preserved the direction of maps in the sequence, while losing exactness on the left. Almost the same behavior occurs if we apply $\text{Hom}_R(N, \bullet)$ to a short exact sequence, except that we lose exactness at the rightmost position. On the other hand, as we saw above, applying $\text{Hom}_R(\bullet, N)$ reverses the direction of the maps:

Exercise 2.2.19 Show that a short exact sequence of R-modules

$$0 \longrightarrow M_2 \xrightarrow{d_2} M_1 \xrightarrow{d_1} M_0 \longrightarrow 0$$

gives rise to a left exact sequence:

$$0 \longrightarrow \mathrm{Hom}_R(M_0, N) \longrightarrow \mathrm{Hom}_R(M_1, N) \longrightarrow \mathrm{Hom}_R(M_2, N)$$

Use the short exact sequence of $\mathbb{Z}$-modules in Exercise 2.2.18 to show that exactness can fail at the rightmost position.   ◇

To represent an element $\phi \in \mathrm{Hom}_R(M_1, M_2)$, we need to account for the fact that modules have both *generators* and *relations*.

Definition 2.2.20 A presentation for an R-module M is a right exact sequence of the form

$$F \xrightarrow{\alpha} G \longrightarrow M \longrightarrow 0,$$

where F and G are free modules. If M is finitely generated, then G can be chosen to have finite rank, so is isomorphic to R^a for $a \in \mathbb{N}$.

Algorithm 2.2.21 *To define a homomorphism between finitely presented R-modules M_1 and M_2, take presentations for M_1 and M_2*

$$R^{a_1} \xrightarrow{\alpha} R^{a_0} \longrightarrow M_1 \longrightarrow 0,$$

and

$$R^{b_1} \xrightarrow{\beta} R^{b_0} \longrightarrow M_2 \longrightarrow 0.$$

An element of $\mathrm{Hom}_R(M_1, M_2)$ is determined by a map $R^{a_0} \xrightarrow{\gamma} R^{b_0}$ which preserves the relations. So if $b = \alpha(a)$, then $\gamma(b) = \beta(c)$. In particular, the image of the composite map

$$R^{a_1} \xrightarrow{\gamma \cdot \alpha} R^{b_0}$$

must be contained in the image of β.

Exercise 2.2.22 For R–modules M, N, and P, prove that

$$\mathrm{Hom}_R(M \otimes_R N, P) \simeq \mathrm{Hom}_R(M, \mathrm{Hom}_R(N, P)),$$

as follows: let

$$\phi \in \mathrm{Hom}_R(M \otimes_R N, P).$$

Given $m \in M$, we must produce an element of $\mathrm{Hom}_R(N, P)$. Since $\phi(m \otimes \bullet)$ takes elements of N as input and returns elements of P as output, it suffices to show that $\phi(m \otimes \bullet)$ is a homomorphism of R-modules, and in fact an isomorphism.   ◇

2.3 Localization of Rings and Modules

The process of quotienting an object M by a subobject N has the effect of making N equal to zero. Localization simplifies a module or ring in a different way, by making some subset of objects invertible.

Definition 2.3.1 Let R be a ring, and S a multiplicatively closed subset of R containing 1. Define an equivalence relation on $\left\{ \frac{a}{b} \mid a \in R, b \in S \right\}$ via

$$\frac{a}{b} \sim \frac{c}{d} \text{ if } (ad - bc)u = 0 \text{ for some } u \in S.$$

Then the localization of R at S is

$$R_S = \left\{ \frac{a}{b} \mid a \in R, b \in S \right\} / \sim .$$

The object R_S is a ring, with operations defined exactly as we expect:

$$\frac{a}{b} \cdot \frac{a'}{b'} = \frac{a \cdot a'}{b \cdot b'} \text{ and } \frac{a}{b} + \frac{a'}{b'} = \frac{a \cdot b' + b \cdot a'}{b \cdot b'}.$$

Exercise 2.3.2 Let $R = \mathbb{Z}$, and let S consist of all nonzero elements of R. Prove that the localization $R_S = \mathbb{Q}$. More generally, prove that if R is a domain and S is the set of nonzero elements of R, then R_S is the field of fractions of R. ◇

The most frequently encountered situations for localization are when

- S is the complement of a prime ideal.
- $S = \{1, f, f^2, \ldots\}$ for some $f \in R$.

When S is the complement of a prime ideal P, it is usual to write R_P for R_S. By construction, everything outside the ideal PR_P is a unit, so R_P has a unique maximal ideal and is thus a local ring. If M is an R-module, then we can construct an R_S-module M_S in the same way R_S was constructed. One reason that localization is a useful tool is that it preserves exact sequences.

Theorem 2.3.3 *Localization preserves exact sequences.*

Proof First, suppose we have a map of R-modules $M \xrightarrow{\phi} M'$. Since ϕ is R-linear, this gives us a map $M_S \xrightarrow{\phi_S} M'_S$ via $\phi_S(\frac{m}{s}) = \frac{\phi(m)}{s}$. Let

$$0 \longrightarrow M' \xrightarrow{\phi} M \xrightarrow{\psi} M'' \longrightarrow 0$$

be an exact sequence. Then

$$\psi_S \phi_S \left(\frac{m}{s} \right) = \frac{\psi(\phi(m))}{s} = 0.$$

On the other hand, if $\frac{m}{s'} \in \ker \psi_S$, so that $\frac{\psi(m)}{s'} \sim \frac{0}{s''}$, then there is an $s \in S$ such that $s\psi(m) = 0$ in R. But $s\psi(m) = \psi(sm)$ so $sm \in \ker \psi = \mathrm{im}\,\phi$, and $sm = \phi(n)$ for some $n \in M'$. Thus, we have $m = \frac{\phi(n)}{s}$ and so $\frac{m}{s'} = \frac{\phi(n)}{ss'}$.   □

Exercise 2.3.4 Let M be a finitely generated R-module, and S a multiplicatively closed set. Show that $M_S = 0$ iff there exists $s \in S$ such that $s \cdot M = 0$.   ◇

Example 2.3.5 For the ideal $I = \langle xy, xz \rangle \subseteq \mathbb{K}[x, y, z] = R$, the quotient R/I is both a ring and an R-module. It is easy to see that

$$I = \langle x \rangle \cap \langle y, z \rangle$$

and by Theorem 2.2.11, $P_1 = \langle x \rangle$ and $P_2 = \langle y, z \rangle$ are prime ideals. What is the effect of localization? Notice that $x \notin P_2$ and $y, z \notin P_1$. In the localization $(R/I)_{P_2}$, x is a unit, so

$$I_{P_2} \simeq \langle y, z \rangle_{P_2}, \text{ so } (R/I)_{P_2} \simeq R_{P_2}/I_{P_2} \simeq \mathbb{K}(x),$$

where $\mathbb{K}(x) = \{\frac{f(x)}{g(x)} \mid g(x) \neq 0\}$. In similar fashion, in $(R/I)_{P_1}$, y is a unit, so

$$I_{P_1} \simeq \langle x \rangle_{P_1}, \text{ so } (R/I)_{P_1} \simeq \mathbb{K}(y, z).$$

Finally, if P is a prime ideal which does not contain I, then there is an element $f \in I \setminus P$. But then

- f is a unit because it is outside P
- f is zero because it is inside I.

Hence

$$(R/I)_P = 0 \text{ if } I \nsubseteq P.$$

Exercise 2.3.6 Carry out the same computation for $I = \langle x^2, xy \rangle \subseteq \mathbb{K}[x, y] = R$. You may find it useful to use the fact that

$$I = \langle x^2, y \rangle \cap \langle x \rangle$$

Hint: $\langle x^2, y \rangle$ is not a prime ideal, but $\langle x, y \rangle$ is prime.   ◇

2.4 Noetherian Rings, Hilbert Basis Theorem, Varieties

Exercise 2.2.13 defined a principal ideal domain; the ring $\mathbb{K}[x]$ of polynomials in one variable with coefficients in a field is an example. In particular, every ideal $I \subseteq \mathbb{K}[x]$ can be generated by a single element, hence the question of when $f(x) \in I$ is easy to solve: find the generator $g(x)$ for I, and check if $g(x)|f(x)$. While this is easy in the univariate case, in general the *ideal membership problem* is difficult. The class of Noetherian rings includes all principal ideal domains, but is much larger; in a Noetherian ring every ideal is finitely generated. Gröbner bases and the Buchberger algorithm are analogs of Gaussian Elimination for polynomials of degree larger than one, and provide a computational approach to tackle the ideal membership question. For details on this, see [47].

2.4.1 Noetherian Rings

Definition 2.4.1 A ring is *Noetherian* if it contains no infinite ascending chains of ideals: there is no infinite chain of proper inclusions of ideals as below

$$I_1 \subsetneq I_2 \subsetneq I_3 \subsetneq \cdots$$

A module is Noetherian if it contains no infinite ascending chains of submodules. A ring is Noetherian exactly when all ideals are finitely generated.

Theorem 2.4.2 *A ring R is Noetherian iff every ideal is finitely generated.*

Proof Suppose every ideal in R is finitely generated, but there is an infinite ascending chain of ideals:

$$I_1 \subsetneq I_2 \subsetneq I_3 \subsetneq \cdots$$

Let $J = \bigcup_{i=1}^{\infty} I_i$. Since $j_1 \in J$, $j_2 \in J$ and $r \in R$ implies $j_1 + j_2 \in J$ and $r \cdot j_i \in J$, J is an ideal. By assumption, J is finitely generated, say by $\{f_1, \ldots, f_k\}$, and each $f_i \in I_{l_i}$ for some l_i. So if $m = \max\{l_i\}$ is the largest index, we have

$$I_{m-1} \subsetneq I_m = I_{m+1} = \cdots,$$

a contradiction. Now suppose that I cannot be finitely generated, so we can find a sequence of elements $\{f_1, f_2, \ldots\}$ of I with $f_i \notin \langle f_1, f_2, \ldots, f_{i-1} \rangle$. This yields

$$\langle f_1 \rangle \subsetneq \langle f_1, f_2 \rangle \subsetneq \langle f_1, f_2, f_3 \rangle \subsetneq \cdots,$$

which is an infinite ascending chain of ideals. $\square$

Exercise 2.4.3 Let M be a module. Prove the following are equivalent:

(a) M contains no infinite ascending chains of submodules.
(b) Every submodule of M is finitely generated.
(c) Every nonempty subset Σ of submodules of M has a maximal element with respect to inclusion.

The last condition says that Σ is a special type of partially ordered set. $\Diamond$

Exercise 2.4.4 Prove that if R is Noetherian and M is a finitely generated R-module, then M is Noetherian, as follows. Since M is finitely generated, there exists an n such that R^n surjects onto M. Suppose there is an infinite ascending chain of submodules of M, and consider what this would imply for R. $\Diamond$

Theorem 2.4.5 [*Hilbert Basis Theorem*] *If R is a Noetherian ring, then so is $R[x]$.*

Proof Let I be an ideal in $R[x]$. By Theorem 2.4.2 we must show that I is finitely generated. The set of lead coefficients of polynomials in I generates an ideal I' of R, which is finitely generated, because R is Noetherian. Let $I' = \langle g_1, \ldots, g_k \rangle$. For each g_i there is a polynomial

$$f_i \in I, \ f_i = g_i x^{m_i} + \text{ terms of lower degree in } x.$$

Let $m = \max\{m_i\}$, and let I'' be the ideal generated by the f_i. Given any $f \in I$, reduce it modulo members of I'' until the lead term has degree less than m. The R-*module* M generated by $\{1, x, \ldots, x^{m-1}\}$ is finitely generated, hence Noetherian. Therefore the submodule $M \cap I$ is also Noetherian, with generators $\{h_1, \ldots, h_j\}$. Hence I is generated by $\{h_1, \ldots, h_j, g_1, \ldots, g_k\}$, which is a finite set. $\square$

When a ring R is Noetherian, even for an ideal $I \subseteq R$ specified by an infinite set of generators, there will exist a finite generating set for I. A field $\mathbb{K}$ is Noetherian, so the Hilbert Basis Theorem and induction tell us that the ring $\mathbb{K}[x_1, \ldots, x_n]$ is Noetherian, as is a polynomial ring over $\mathbb{Z}$ or any other principal ideal domain. Thus, the goal of determining a finite generating set for an ideal is attainable.

2.4.2   Solutions to a Polynomial System: Varieties

In linear algebra, the objects of study are the solutions to systems of polynomial equations of the simplest type: all polynomials are of degree one. Algebraic geometry is the study of the sets of solutions to systems of polynomial equations of higher degree. The choice of field is important: $x^2 + 1 = 0$ has no solutions in $\mathbb{R}$ and two solutions in $\mathbb{C}$; in linear algebra this issue arose when computing eigenvalues of a matrix.

Definition 2.4.6 A field $\mathbb{K}$ is *algebraically closed* if every nonconstant polynomial $f(x) \in \mathbb{K}[x]$ has a solution $f(p) = 0$ with $p \in \mathbb{K}$. The *algebraic closure* $\overline{\mathbb{K}}$ is the smallest field containing $\mathbb{K}$ which is algebraically closed.

Given a system of polynomial equations $\{f_1, \ldots, f_k\} \subseteq R = \mathbb{K}[x_1, \ldots, x_n]$, note that the set of common solutions over $\mathbb{K}$ depends only on the ideal

$$I = \langle f_1, \ldots, f_k \rangle = \{\sum_{i=1}^{k} g_i f_i \mid g_i \in R\}$$

The set of common solutions is called the *variety* of I, denoted

$$\mathbf{V}(I) \subseteq \mathbb{K}^n \subseteq \overline{\mathbb{K}}^n$$

Adding more equations to a polynomial system imposes additional constraints on the solutions, hence passing to varieties reverses inclusion

$$I \subseteq J \Rightarrow \mathbf{V}(J) \subseteq \mathbf{V}(I)$$

Since ideals $I, J \subseteq R$ are submodules of the same ambient module, we have

- $I \cap J = \{f \mid f \in I \text{ and } f \in J\}$.
- $IJ = \langle fg \mid f \in I \text{ and } g \in J \rangle$.
- $I + J = \{f + g \mid f \in I \text{ and } g \in J\}$.

It is easy to check these are all ideals.

Exercise 2.4.7 Prove that

$$\mathbf{V}(I \cap J) = \mathbf{V}(I) \cup \mathbf{V}(J) = \mathbf{V}(IJ)$$

and that $\mathbf{V}(I + J) = \mathbf{V}(I) \cap \mathbf{V}(J)$.   ◇

Definition 2.4.8 The radical of an ideal $I \subseteq R$ is

$$\sqrt{I} = \{f \mid f^m \in I \text{ for some power m}\}.$$

Exercise 2.4.9 Show if $I = \langle x^2 - xz, xy - yz, xz - z^2, xy - xz, y^2 - yz, yz - z^2 \rangle$ in $\mathbb{K}[x, y, z]$, then $\sqrt{I} = \langle x - z, y - z \rangle$.   ◇

Definition 2.4.10 For ideals $I, J \subseteq R$, the ideal quotient (or colon ideal) is

$$I : J = \{f \in R \mid f \cdot J \subseteq I\}.$$

Exercise 2.4.11 For I and $\sqrt{I}$ as in Exercise 2.4.9 show $I : \sqrt{I} = \langle x, y, z \rangle$.   ◇

The operations of radical and ideal quotient have geometric interpretations; we tackle the radical first. Given $X \subseteq \mathbb{K}^n$, consider the set $I(X)$ of all polynomials in $R = \mathbb{K}[x_1, \ldots, x_n]$ which vanish on X.

Exercise 2.4.12 Prove that $I(X)$ is an ideal, and in fact a radical ideal. Next, prove that $\mathbf{V}(I(X))$ is the smallest variety containing X.   ◇

Notice that for $I = \langle x^2 + 1 \rangle \subseteq \mathbb{R}[x] \subseteq \mathbb{C}[x]$, $\mathbf{V}(I)$ is empty in $\mathbb{R}$, but consists of two points in $\mathbb{C}$. This brings us Hilbert's *Nullstellensatz* (see [70] for a proof).

Theorem 2.4.13 *If* $\mathbb{K}$ *is algebraically closed, then*

- *Hilbert Nullstellensatz (Weak version):* $\mathbf{V}(I) = \emptyset \iff 1 \in I$.
- *Hilbert Nullstellensatz (Strong version):* $I(\mathbf{V}(I)) = \sqrt{I}$.

An equivalent formulation is that over an algebraically closed field, there is a $1:1$ correspondence between maximal ideals $I(p)$ and points p. For an arbitrary set $X \subseteq \mathbb{K}^n$, the smallest variety containing X is $\mathbf{V}(I(X))$. It is possible to define a topology, called the *Zariski topology* on $\mathbb{K}^n$, where the closed sets are of the form $\mathbf{V}(I)$, and when working with this topology we write $\mathbb{A}_{\mathbb{K}}^n$ and speak of *affine space*. We touch on the Zariski topology briefly in the general discussion of topological spaces in Chap. 3. We close with the geometry of the ideal quotient operation. Suppose we know there is some spurious or unwanted component $\mathbf{V}(J) \subseteq \mathbf{V}(I)$. How do we remove $\mathbf{V}(J)$? Equivalently, what is the smallest variety containing $\mathbf{V}(I) \setminus \mathbf{V}(J)$?

Theorem 2.4.14 *The variety* $\mathbf{V}(I(\mathbf{V}(I) \setminus \mathbf{V}(J))) \subseteq \mathbf{V}(I : J)$.

Proof Since $I_1 \subseteq I_2 \Rightarrow \mathbf{V}(I_2) \subseteq \mathbf{V}(I_1)$, it suffices to show

$$I : J \subseteq I(\mathbf{V}(I) \setminus \mathbf{V}(J)).$$

If $f \in I : J$ and $p \in \mathbf{V}(I) \setminus \mathbf{V}(J)$, then since $p \notin \mathbf{V}(J)$ there is a $g \in J$ with $g(p) \neq 0$. Since $f \in I : J$, $fg \in I$, and so

$$p \in \mathbf{V}(I) \Rightarrow f(p)g(p) = 0.$$

As $g(p) \neq 0$, this forces $f(p) = 0$ and therefore $f \in I(\mathbf{V}(I) \setminus \mathbf{V}(J))$.   □

Example 2.4.15 Projective space $\mathbb{P}_{\mathbb{K}}^n$ over a field $\mathbb{K}$ is defined as

$$\mathbb{K}^{n+1} \setminus \{0\} / \sim \text{ where } p_1 \sim p_2 \text{ iff } p_1 = \lambda p_2 \text{ for some } \lambda \in \mathbb{K}^*.$$

One way to visualize $\mathbb{P}_{\mathbb{K}}^n$ is as the set of lines through the origin in $\mathbb{K}^{n+1}$.

Writing $\mathbf{V}(x_0)^c$ for the points where $x_0 \neq 0$, we have that $\mathbb{K}^n \simeq \mathbf{V}(x_0)^c \subseteq \mathbb{P}_{\mathbb{K}}^n$ via $(a_1, \ldots, a_n) \mapsto (1, a_1, \ldots, a_n)$. This can be quite useful, since $\mathbb{P}_{\mathbb{K}}^n$ is compact.

Exercise 2.4.16 Show that a polynomial $f = \sum c_{\alpha_i} x^{\alpha_i} \in \mathbb{K}[x_0, \ldots, x_n]$ has a well-defined zero set in $\mathbb{P}^n_{\mathbb{K}}$ iff it is *homogeneous*: the exponents $\alpha_i = (\alpha_{i_0}, \ldots, \alpha_{i_n})$ all have the same weight $\sum_j \alpha_{i_j} = k$ for a fixed k. Put differently, all the α_i have the same dot product with the vector $[1, \ldots, 1]$. Show that a homogeneous polynomial f does not define a function on $\mathbb{P}^n_{\mathbb{K}}$, but that the *rational function* $\frac{f}{g}$ with f and g homogeneous of the same degree does define a function on $\mathbb{P}^n_{\mathbb{K}} \setminus \mathbf{V}(g)$.   ◇

Example 2.4.17 Let I be the ideal of 2×2 minors of

$$\begin{bmatrix} x & y & z \\ y & z & w \end{bmatrix}, \text{ so } I = \langle xz - y^2, xw - yz, yw - z^2 \rangle,$$

and let $J = \langle xz - y^2, xw - yz \rangle$ and $L = \langle x, y \rangle$. Then $\mathbf{V}(I)$ is a curve in $\mathbb{P}^3$:

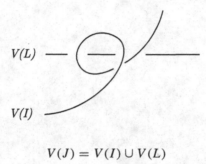

$$V(J) = V(I) \cup V(L)$$

Exercise 2.4.18 Prove the equality above. Hint: use ideal quotient.   ◇

Chapter 3
Basics of Topology: Spaces and Sheaves

In Chap. 2 we covered the basics of upper level algebra. Now we do the same thing for topology: this chapter covers some key ideas of topology and sheaf theory. We start in Sect. 3.1 with a quick review of topological basics. In Sect. 3.2 we discuss vector bundles, where the central construction provides a good intuition for sheaf theory, which is introduced in Sect. 3.3. Sheaf theory can be a somewhat daunting topic when first encountered; the main point to keep in mind is that a sheaf is really nothing more than a bookkeeping device.

- Sheaves organize local data, assigning to every open set U an algebraic object $\mathscr{F}(U)$. When $V \subseteq U$, there is a restriction map

$$\mathscr{F}(U) \xrightarrow{\rho_{UV}} \mathscr{F}(V)$$

  satisfying several natural conditions which appear in Definition 3.3.1.
- Sheaves facilitate the construction of global structures from local structures: this is often referred to as *gluing*, formalized in Definition 3.3.2.

The chapter concludes with a description of the work of Hansen-Ghrist applying sheaf theory to social networks. This requires a quick introduction to spectral graph theory, relating the graph Laplacian to the heat equation. The main idea is that the heat equation is a good model for the spread of opinion over a social media network. The topics covered in this chapter are:

- Topological Spaces.
- Vector Bundles.
- Sheaf Theory.
- Sheaves, Graphs, and Social Media.

© The Author(s), under exclusive license to Springer Nature Switzerland AG 2022   43
H. Schenck, *Algebraic Foundations for Applied Topology and Data Analysis*,
Mathematics of Data 1, https://doi.org/10.1007/978-3-031-06664-1_3

3.1 Topological Spaces

3.1.1 Set Theory and Equivalence Relations

In this section we fix notation, and review basic topology and set theory. A set S is a (not necessarily finite) collection of elements $\{s_1, \ldots, s_k, \ldots\}$; standard operations among sets include

(a) union: $S_1 \cup S_2 = \{m \mid m \in S_1 \text{ or } m \in S_2\}$.
(b) intersection: $S_1 \cap S_2 = \{m \mid m \in S_1 \text{ and } m \in S_2\}$.
(c) set difference: $S_1 \setminus S_2 = \{m \mid m \in S_1 \text{ and } m \notin S_2\}$.

The Cartesian product $S_1 \times S_2$ of two sets is

$$S_1 \times S_2 = \{(s_1, s_2) \mid s_i \in S_i\}$$

and a relation $\sim$ between S_1 and S_2 is a subset of pairs of $S_1 \times S_2$. We say that a relation on $S \times S$ is

(a) reflexive if $a \sim a$ for all $a \in S$.
(b) symmetric if $a \sim b \Rightarrow b \sim a$ for all $a, b \in S$.
(c) transitive if $a \sim b$ and $b \sim c \Rightarrow a \sim c$ for all $a, b, c \in S$.

A relation which is reflexive, symmetric, and transitive is an *equivalence relation*.

Exercise 3.1.1 Show that an equivalence relation partitions a set into disjoint subsets. Each of these subsets (which consist of elements related by $\sim$) is called an *equivalence class*.   ◇

A relation $R \subseteq S \times S$ is a *partial order* on S if it is transitive, and (s_1, s_2) and $(s_2, s_1) \in R$ imply that $s_1 = s_2$. In this case S is called a *poset*, which is shorthand for "partially ordered set".

3.1.2 Definition of a Topology

The usual setting for calculus and elementary analysis is over the real numbers; $\mathbb{R}^n$ is a metric space, with $d(p, q)$ the distance between points p and q. A set $U \subseteq \mathbb{R}^n$ is open when for every $p \in U$ there exists $\delta_p > 0$ such that

$$N_{\delta_p}(p) = \{q \mid d(q, p) < \delta_p\} \subseteq U,$$

and closed when the complement $\mathbb{R}^n \setminus U$ is open. The key properties of open sets are that any (including infinite) union of open sets is open, and a finite intersection of open sets is open. Topology takes this as a cue.

Definition 3.1.2 A *topology* on a set S is a collection $\mathcal{U}$ of subsets $U \subseteq S$, satisfying conditions (a)–(c) below. Elements of $\mathcal{U}$ are the *open sets* of the topology, and $(S, \mathcal{U})$ (or, for short, just S) is a *topological space*.

(a) $\emptyset$ and S are elements of $\mathcal{U}$.
(b) $\mathcal{U}$ is closed under arbitrary union: for $U_i \in \mathcal{U}$, $\bigcup_{i \in I} U_i \in \mathcal{U}$.
(c) $\mathcal{U}$ is closed under finite intersection: for $U_i \in \mathcal{U}$, $\bigcap_{i=1}^{m} U_i \in \mathcal{U}$.

Exercise 3.1.3 Prove that a topology can also be defined by specifying the closed sets, by complementing the conditions of Definition 3.1.2.   ◇

Since $\mathbb{R}^n$ is a metric space, the notions of distance and limit points are intuitive. How can we define limit points in the setting of an abstract topological space?

Definition 3.1.4 If S is a topological space, then $p \in S$ is a *limit point* of $X \subseteq S$ if for all $U \in \mathcal{U}$ with $p \in U$,

$$(U \setminus p) \cap X \neq \emptyset$$

The closure $\overline{X}$ of X is defined as $X \cup \{q \mid q \text{ is a limit point of } X\}$.

Definition 3.1.5 A collection $\mathcal{B}$ of subsets of S is a *basis* for the topology if every finite intersection of elements of $\mathcal{B}$ is a union of elements of $\mathcal{B}$. It is not hard to show that the collection of (arbitrary) unions of elements of $\mathcal{B}$ is a topology, called the topology generated by $\mathcal{B}$.

Example 3.1.6 In $\mathbb{R}^n$, $\mathcal{B} = \{N_\delta(p) \mid p \in \mathbb{R}^n \text{ and } \delta \in \mathbb{R}_{>0}\}$ is a basis.

Our last definitions involve properties of *coverings* and *continuity*:

Definition 3.1.7 A collection $\mathcal{U}$ of subsets of S is a *cover* of S if

$$\bigcup_{U_i \in \mathcal{U}} U_i = S.$$

S is *compact* if every open covering admits a finite subcover, and *connected* if the only sets that are both open and closed are $\emptyset$ and S.

Example 3.1.8 [Heine-Borel] A set $S \subseteq \mathbb{R}^n$ is compact iff it is closed and bounded.

Definition 3.1.9 Let S and T be topological spaces. A function

$$S \xrightarrow{f} T$$

is *continuous* if $f^{-1}(U)$ is open in S for every open set $U \subseteq T$. We say that S and T are *homeomorphic* if f and f^{-1} are 1:1 and continuous for some map f as above. Two maps $f_0, f_1 : S \to T$ are *homotopic* if there is a continuous map

$$S \times [0, 1] \xrightarrow{F} T$$

such that $F(x, 0) = f_0(x)$ and $F(x, 1) = f_1(x)$. F deforms the map f_0 to the map f_1. The spaces S and T are *homotopic* if there is a map $g : T \to S$ such that $f \circ g$ is homotopic to 1_T and $g \circ f$ is homotopic to 1_S. We say that a topological space is *contractible* if it is homotopic to a point.

Exercise 3.1.10 Prove that if $S \xrightarrow{f} T$ is continuous and X is a compact subset of S, then $f(X)$ is a compact subset of T.   ◇

3.1.3   Discrete, Product, and Quotient Topologies

We now discuss some common topologies, as well as how to build new topological spaces from old. Our first example is

Definition 3.1.11 (Discrete Topology) The topology in which every subset A of S is open is called the *discrete topology*. Since the complement $S \setminus A$ is also a subset, this means every subset is both open and closed.

From the perspective of data science, point cloud data only comes equipped with the discrete topology; by adding a parameter we are able to bring interesting topological features into the picture in Chap. 7 with persistent homology.

Definition 3.1.12 (Product Topology) If X_1 and X_2 are topological spaces, then the Cartesian product has a natural topology associated with it, the *product topology*, where open sets are unions of sets of the form $U_1 \times U_2$ with U_i open in X_i. In particular, if B_i is a basis for X_i, then taking the U_i to be members of B_i shows that $B_1 \times B_2$ is a basis for $X_1 \times X_2$.

Exercise 3.1.13 Show that the product topology is the coarsest topology—the topology with the fewest open sets—such that the projection maps

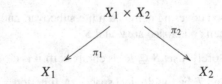

are continuous, where $\pi_i(x_1, x_2) = x_i$.   ◇

Definition 3.1.14 [Induced and Quotient Topology]

• If $Y \xhookrightarrow{i} X$ then the *induced topology* on Y is the topology where open sets of Y are $i^{-1}(U)$ for U open in X.

- Let X be a topological space and $\sim$ an equivalence relation. The quotient space $Y = X/\!\!\sim$ is the set of equivalence classes, endowed with the *quotient topology*: $U \subseteq Y$ is open iff

$$V = \pi^{-1}(U) \text{ is open in } X, \text{ where } X \xrightarrow{\pi} X/\!\!\sim .$$

Example 3.1.15 For $W \subseteq X$, define an equivalence relation via

$$a \sim b \Longleftrightarrow a, b \in W \text{ or } a = b.$$

This collapses W to a point, so for example if $X = [0, 1]$ with the induced topology from $\mathbb{R}^1$ and W consists only of the two endpoints $\{0, 1\}$ then X/W is homeomorphic to S^1, and if X is the unit square $[0, 1] \times [0, 1]$ and $W = \partial(X)$ then X/Y is homeomorphic to S^2. In contrast, if we define an equivalence relation on X via $(x, 1) \sim (x, 0)$ and $(0, y) \sim (1, y)$ then $X/\!\!\sim$ is homeomorphic to the torus T^2.

A quintessential example of a topological space is a *differentiable manifold*. As the name indicates, there is an additional geometric (differentiable) structure involved; the basic intuition is that a differentiable manifold of dimension n "is locally the same as" Euclidean space $\mathbb{R}^n$. Consider the unit circle $V(x^2 + y^2 - 1) \subset \mathbb{R}^2$. In this case, the circle inherits the induced topology from $\mathbb{R}^2$, and differentiable structure from the differentiable structure of $\mathbb{R}^2$.

But not every topological space comes to us embedded in $\mathbb{R}^n$. A famous result due to Hassler Whitney (which we shall not prove) is that every n-dimensional manifold can be embedded in $\mathbb{R}^{2n}$. We start by defining more precisely what we mean by "is locally the same as". A map $V \xrightarrow{f} V'$ between two open subsets of $\mathbb{R}^n$ is a *diffeomorphism* if it is $1 : 1$, onto, and has continuous derivatives of all orders (which is also written as $f \in C^\infty$), as does f^{-1}.

Definition 3.1.16 Let I be an index set, and consider a collection of open sets

$$\mathscr{U} = \{U_i, i \in I\} \text{ with each } U_i \subset \mathbb{R}^n.$$

A *chart* is a *diffeomorphism* ϕ_i from an open set $U_i \longrightarrow \mathbb{R}^m$, and an *atlas* is a collection of charts for all $i \in I$. We require that the ϕ_i satisfy a compatibility condition: $\phi_i = \phi_j$ on $U_i \cap U_j$. An n-dimensional differentiable manifold X is constructed by identifying the $\phi_i(U_i)$ on common intersections, and applying Definition 3.1.14.

The key constructions in calculus can then be carried over to this setting.

Example 3.1.17 Let

$$U_0 = (-\epsilon, \pi + \epsilon) = U_1 \subseteq \mathbb{R}^1,$$

· and let $U_i \xrightarrow{\phi_i} \mathbb{R}^2$ via

$$\phi_i(x) = ((-1)^i \cos(x), \sin(x))$$

So U_0 is mapped to the top half (plus ϵ) of the unit circle, and U_1 to the bottom half (plus ϵ), with $U_0 \cap U_1$ equal to the two small open segments.

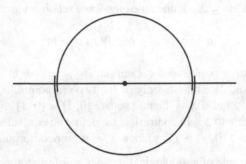

This example shows that Whitney's result is tight–the circle is one dimensional, but cannot be embedded in $\mathbb{R}^1$. Building complicated structures by gluing together simple building blocks is the focus of the next section, on vector bundles.

3.2 Vector Bundles

While vector bundles are fundamental objects in topology, the reason to include them in these notes is that they provide the right intuition to motivate the definition of a sheaf. In this section we work with real, *topological* vector bundles, where maps between objects are continuous, written as C^0; for algebraic vector bundles see [48]. The formal definition appears below; the intuition is that a rank n vector bundle $\mathscr{V}$ on a space X is locally a product space of the form

$$U_i \times \mathbb{R}^n, \text{ with } U_i \in \mathscr{U} \text{ a cover of } X,$$

along with projection maps $\mathscr{V} \to U_i$. We will also touch briefly on geometry to discuss the tangent and cotangent bundles of a manifold, where the maps between objects will be differentiable, written as C^∞.

Example 3.2.1 A product space of the form $X \cong Y \times \mathbb{R}^n$ (where $\cong$ denotes homeomorphism) is a *trivial* vector bundle; for example an infinite cylinder C is

$$C \cong S^1 \times \mathbb{R}^1$$

A small connected open set $U \subseteq S^1$ is homeomorphic to $\mathbb{R}^1$, and C can be covered by sets of this form. On the other hand, the Möbius band M is also a rank one vector bundle on S^1, and locally has the same form as C. As Exercise 3.2.5 shows, the spaces M and C are different.

Definition 3.2.2 A topological space $\mathcal{V}$ is a rank n vector bundle (typically called the *total space*) on a topological space X (typically called the *base space*) if there is a continuous map

$$\mathcal{V} \xrightarrow{\ \pi\ } X$$

such that

- There exists an open cover $\mathcal{U} = \{U_i\}_{i \in I}$ of X, and homeomorphisms

$$\pi^{-1}(U_i) \xrightarrow{\ \phi_i\ } U_i \times \mathbb{R}^n$$

such that ϕ_i followed by projection onto U_i is π, and
- The transition maps $\phi_{ij} = \phi_i \circ \phi_j^{-1}$:

$$\phi_j \pi^{-1}(U_j) \supseteq (U_i \cap U_j) \times \mathbb{R}^n \xrightarrow{\ \phi_{ij}\ } (U_i \cap U_j) \times \mathbb{R}^n \subseteq \phi_i \pi^{-1}(U_i)$$

satisfy $\phi_{ij} \in GL_n(C^0(U_i \cap U_j))$.

A cover $\{U_i\}$ so that the first bullet above holds is a *trivialization*, because locally a vector bundle on X looks like a product of U_i with $\mathbb{R}^n$. The map ϕ_j gives us a *chart*: a way to think of a point in $\pi^{-1}(U_j)$ as a point $(t, v) \in U_j \times \mathbb{R}^n$. The copy of $\mathbb{R}^n \cong \phi_i \pi^{-1}(p)$ over a point p is called the *fiber* over p. If we change from a chart over U_j to a chart over U_i, then (Fig. 3.1)

$$(t, v_j) \mapsto (t, [\phi_{ij}(t)] \cdot v_j)$$

We've seen this construction in the first chapter, in the section on change of basis. Suppose B_1 and B_2 are different bases for a vector space V, with $[v]_{B_i}$ denoting the representation of $v \in V$ with respect to basis B_i. Then if Δ_{ij} denotes the transition matrix from B_j to B_i, we have

$$[v]_{B_2} = \Delta_{21}[v]_{B_1}$$

The ϕ_{ij} are *transition functions*, and are a *family* of transition matrices, which vary as $p \in U_i$ varies. A vector space does not have a canonical basis, and this is therefore also true for the fibers of a vector bundle. It follows from the definition that $\phi_{ik} = \phi_{ij} \circ \phi_{jk}$ on $U_i \cap U_j \cap U_k$ and that ϕ_{ii} is the identity.

Fig. 3.1 Visualizing a vector bundle

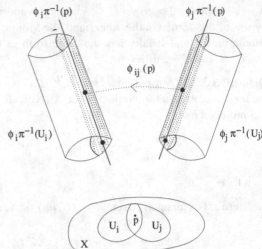

Fig. 3.2 For a section s, $s(p) \in \phi_i \pi^{-1}(p)$, the fiber over p

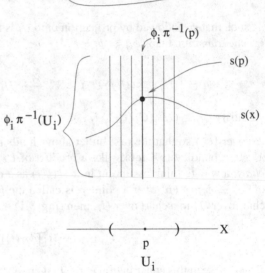

Definition 3.2.3 A *section* of a vector bundle $\mathscr{V}$ over an open set $U \subseteq X$ is a continuous map $U \xrightarrow{s} \mathscr{V}$ such that $\pi \circ s$ is the identity (Fig. 3.2).

Let $\mathscr{S}(U)$ denote the set of all sections of $\mathscr{V}$ over U. A projection $U \to \mathbb{R}^1$ is given by a continuous function $f \in C^0(U)$, so a map from U to $\mathbb{R}^n$ is given by an n–tuple of continuous functions, and is therefore an element of $C^0(U)^n$. This means that locally $\mathscr{S}(U)$ has the structure of a free module over the ring $C^0(U)$ of continuous functions on U.

Exercise 3.2.4 Use properties of the transition functions and continuity to show that if

$$s_j = (f_1, \ldots, f_n) \in \mathscr{S}(U_j) \text{ and } s_i = (g_1, \ldots, g_n) \in \mathscr{S}(U_i)$$

are sections such that

$$\phi_{ij}(s_j|_{U_i \cap U_j}) = s_i|_{U_i \cap U_j}.$$

then there exists $s \in \mathscr{S}(U_i \cup U_j)$ which restricts to the given sections. When $\mathscr{S}(U_i)$ is a free module over some ring of functions on all sufficiently small open sets U_i, then $\mathscr{S}$ is called a *locally free sheaf*, discussed in the next section.   ◇

Exercise 3.2.5 Let $S \subseteq \mathbb{R}^3$ be a smooth surface, $p \in S$ and $p(t), t \in [0, 1]$ a smooth loop on S with $p(0) = p = p(1)$. If $\mathbf{n}(p(0))$ is a normal vector to S at p and $p(t)$ goes through charts $U_i, i \in \{1, \ldots, n\}$ and $\phi_{i+1,i}$ are the transition functions from U_i to U_{i+1}, then S is *orientable* if $\phi_{1n} \cdots \phi_{21}\mathbf{n}(p(0)) = \mathbf{n}(p(1))$. Show a cylinder is orientable but the Möbius strip is not.   ◇

3.3 Sheaf Theory

We introduced vector bundles in the previous section because they give a good intuition for sheaves. For vector bundles, on any open set the fibers over the base are the same; this is not the case for an arbitrary sheaf. But the key concept is the same: a sheaf is defined by local data, together with information that stipulates how to transition from one coordinate system (known as a chart) to another.

3.3.1 Presheaves and Sheaves

Definition 3.3.1 Let X be a topological space. A *presheaf* $\mathscr{F}$ on X is a rule associating to each open set $U \subseteq X$ some type of algebraic object $\mathscr{F}(U)$:

$$U \longrightarrow \mathscr{F}(U),$$

together with the conditions below

(a) If $U \subseteq V$ there is a homomorphism $\mathscr{F}(V) \xrightarrow{\rho_{UV}} \mathscr{F}(U)$. Note this is the opposite of the notation used in [90], §II.1.
(b) $\mathscr{F}(\emptyset) = 0$.

(c) $\rho_{UU} = id_{\mathscr{F}(U)}$.

(d) If $U \subseteq V \subseteq W$ then $\rho_{UV} \circ \rho_{VW} = \rho_{UW}$.

The map ρ_{UV} is called *restriction*. For $s \in \mathscr{F}(V)$ we write $\rho_{UV}(s) = s|_U$.

A sheaf is a presheaf which satisfies two additional properties.

Definition 3.3.2 A presheaf $\mathscr{F}$ on X is a *sheaf* if for any open set $U \subseteq X$ and open cover $\{V_i\}$ of U,

(a) If $s \in \mathscr{F}(U)$ satisfies $s|_{V_i} = 0$ for all i, then $s = 0$ in $\mathscr{F}(U)$.

(b) For $v_i \in \mathscr{F}(V_i)$ and $v_j \in \mathscr{F}(V_j)$ such that $v_i|_{(V_i \cap V_j)} = v_j|_{(V_i \cap V_j)}$, there exists $t \in \mathscr{F}(V_i \cup V_j)$ such that $t|_{V_i} = v_i$ and $t|_{V_j} = v_j$.

The second condition is referred to as a *gluing*: when elements on sets U_1 and U_2 agree on $U_1 \cap U_2$, they can be glued to give an element on $U_1 \cup U_2$.

Example 3.3.3 A main source of examples comes from functions on X. When X has more structure, for example when X is a real or complex manifold, then the choice of functions will typically reflect this structure.

X	$\mathscr{F}(U)$
topological space	C^0 functions on U.
smooth real manifold	C^∞ functions on U.
complex manifold	holomorphic functions on U.

The $\mathscr{F}(U)$ above are all rings, which means in any of those situations it is possible to construct sheaves $\mathscr{M}$, where $\mathscr{M}(U)$ is a $\mathscr{F}(U)$-module. Notice how this fits with Exercise 3.2.4 on the sections of a topological vector bundle.

Example 3.3.4 Let X be a manifold as in Definition 3.1.16, and let $C^\infty(U)$ be the ring of C^∞ functions on an open set $U \subset X$. The *module of $\mathbb{R}$-derivations* $\text{Der}(C^\infty(U))$ is the set of $\mathbb{R}$-linear maps $d : C^\infty(U) \to C^\infty(U)$ such that

$$d(f_1 f_2) = f_1 d(f_2) + f_2 d(f_1).$$

Let $\mathfrak{m}_p$ be the ideal of $C^\infty(U)$ functions vanishing at $p \in X$. Then $C^\infty(U)/\mathfrak{m}_p$ is isomorphic to $\mathbb{R}$, and $\mathbb{R} \otimes \text{Der}(C^\infty(U)_{\mathfrak{m}_p})$ is the tangent space $T_p(X)$. The *tangent sheaf* $\mathscr{T}_X$ is defined by "bundling together" the spaces $T_p(X)$. Chapter 5 introduces the sheaf of one-forms Ω_X^1, which is dual to $\mathscr{T}_X$.

One of the earliest uses of sheaf theory was in complex analysis: the Mittag-Leffler problem involves constructing a meromorphic function with residue 1 at all integers. Clearly in some small neighborhood of $k \in \mathbb{Z}$ the function $\frac{1}{z-k}$ works, so the question is how to glue things together. Let C be a one dimensional connected compact complex manifold–so C is topologically a smooth compact real surface of genus g. On a small open subset corresponding to a complex disk, there are plenty of holomorphic functions. Exercise 3.3.5 shows that trying to glue them together to give a global nonconstant holomorphic function is doomed to failure.

Exercise 3.3.5 The maximum modulus principle in complex analysis is that if $f(z)$ is holomorphic on a bounded domain D, then $|f(z)|$ attains its maximum on $\partial(D)$. Prove there are no nonconstant holomorphic functions on C as above. ◇

3.3.2 Posets, Direct Limit, and Stalks

Sheaves carry local data. The previous discussion involved the complex numbers, where the topological structure arises automatically from the fact that $\mathbb{C}$ is a metric space. For $p \in \mathbb{C}$, functions f and g have the same germ at p when their Taylor series expansions at p agree, which allows a gluing construction. In the setting of persistent and multiparameter persistent homology, the sheaves we encounter will involve the polynomial ring $R = \mathbb{K}[x_1, \ldots, x_n]$ over a field $\mathbb{K}$. The fundamental topology used in this context is the Zariski topology.

Definition 3.3.6 The *Zariski Topology* on $\mathbb{K}^n$ has as closed sets $\mathbf{V}(I)$, ranging over all ideals $I \subseteq R$. Clearly $\mathbf{V}(1) = \emptyset$ and $\mathbf{V}(0) = \mathbb{K}^n$. The corresponding topological space is known as *affine space*, written as $\mathbb{A}_{\mathbb{K}}^n$ or Spec(R). As $\mathbf{V}(I) \subseteq \mathbb{A}_{\mathbb{K}}^n$, it inherits the induced topology appearing in Definition 3.1.14.

Exercise 3.3.7 Use Exercise 2.4.7 of Chap. 2 to show that Definition 3.3.6 does define a topology. A topological space is *Noetherian* if there are no infinite descending chains of closed sets. Prove the Zariski topology is Noetherian. ◇

In the classical topology on $\mathbb{R}^n$, the notion of smaller (and smaller) ϵ-neighborhoods of a point is clear and intuitive. In the Zariski topology, all open sets are complements of varieties, so are dense in the classical topology—the open sets are *very big*. What is the correct analog of a *small open set* in this context? The solution is a formal algebraic construction, known as the direct limit.

Definition 3.3.8 A *directed set* S is a partially ordered set with the property that if $i, j \in S$ then there exists $k \in S$ with $i \leq k$, $j \leq k$. Let R be a ring and $\{M_i\}$ a collection of R-modules, indexed by a directed set S, such that for each pair $i \leq j$ there exists a homomorphism $\mu_{ji} : M_i \to M_j$. If $\mu_{ii} = 1_{M_i}$ for all i and $\mu_{kj} \circ \mu_{ji} = \mu_{ki}$ for all $i \leq j \leq k$, then the modules M_i are said to form a *directed system*. Given a directed system, the *direct limit* is an R-module constructed as follows: let N be the submodule of $\oplus M_l$ generated by the relations $m_i - \mu_{ji}(m_i)$, for $m_i \in M_i$ and $i \leq j$. Then the direct limit is

$$\varinjlim_{S} M_i = (\bigoplus_{i \in S} M_i)/N.$$

To phrase it a bit differently: elements $m_i \in M_i$ and $m_j \in M_j$ are identified in the direct limit if the images of m_i and m_j eventually agree.

Definition 3.3.9 The *stalk* of a sheaf at a point is the direct limit over all open sets U containing p:

$$\mathscr{F}_p = \varinjlim_{p \in U} \mathscr{F}(U).$$

The stalk is obtained by taking all open neighborhoods U_i of p, and then using the restriction maps to identify elements which agree on a small enough neighborhood. This mirrors the construction of the germ of a function discussed above: if $\mathscr{H}(U)$ is the sheaf of holomorphic functions on $U \subseteq \mathbb{C}$ in the classical topology, then given

$$f_1 \in \mathscr{H}(U_1) \text{ and } f_2 \in \mathscr{H}(U_2)$$

such that $p \in U_i$, then $f_1 = f_2$ in the stalk $\mathscr{H}_p$ if they have the same Taylor series expansion at p. The stalk at p consists of convergent power series at p. In particular, the stalk encodes very local information. Because stalks represent very local data, they serve as the building blocks for algebraic constructions with sheaves.

3.3.3 Morphisms of Sheaves and Exactness

When a sheaf is given locally by an algebraic object, standard algebraic constructions will have sheaf-theoretic analogs. However, there are usually some subtle points to deal with, as illustrated below.

Definition 3.3.10 A morphism of sheaves

$$\mathscr{F} \xrightarrow{\phi} \mathscr{G}$$

on X is defined by giving, for all open sets U, maps

$$\mathscr{F}(U) \xrightarrow{\phi(U)} \mathscr{G}(U),$$

which commute with the restriction maps.

- ϕ is injective if for any $p \in X$ there exists an open set U with $\phi(U)$ injective (which implies $\phi(V)$ is injective for all $V \subseteq U$).
- ϕ is surjective if for any $p \in X$ and $g \in \mathscr{G}(U)$ there exists $V \subseteq U$ with $p \in V$ and $f \in \mathscr{F}(V)$ such that $\phi(V)(f) = \rho_{UV}(g)$.

For a morphism of sheaves as above, the assignment $U \mapsto \ker(\phi)(U)$ defines a sheaf. However, the image and cokernel of ϕ are only presheaves, and it is necessary to do a bit of preparatory work, by building a *sheaf associated to a presheaf*, to obtain the sheaf-theoretic version of image and cokernel. A sequence of sheaves and sheaf homomorphisms

$$\cdots \longrightarrow \mathscr{F}_{i+1} \xrightarrow{d_{i+1}} \mathscr{F}_i \xrightarrow{d_i} \mathscr{F}_{i-1} \longrightarrow \cdots$$

is exact when the kernel sheaf of d_i is equal to the image sheaf of d_{i+1}. For proofs of the theorem below, as well as for the construction of the sheaf associated to a presheaf mentioned above, see Chapter II of [90].

Theorem 3.3.11 *A sequence of sheaves is exact iff it is exact at the level of stalks.*

At the start of this chapter, we noted that sheaves are bookkeeping devices which encode when objects glue together, as well as when there are obstructions. The *global sections* of $\mathscr{F}$ are globally defined elements of $\mathscr{F}$, and are written as $\mathscr{F}(X)$ or $\Gamma(\mathscr{F})$ or $\check{H}^0(\mathscr{F})$. In Chap. 5 we define Čech cohomology, which is a homology theory specifically designed to measure the conditions (and obstructions) involved in gluing local sections together.

3.4 From Graphs to Social Media to Sheaves

This section highlights recent work of Hansen–Ghrist [86], [87] connecting sheaf theory to graphs and social media. We begin with a quick introduction to *spectral graph theory*, focusing on the graph Laplacian. The graph Laplacian can be thought of as encoding solutions of the heat equation (Chapter 14 of [139])

$$\frac{\partial u}{\partial t} = -k\Delta u = -k\frac{\partial^2 u}{\partial x^2}, \tag{3.4.1}$$

where $u = u(x, t)$ is temperature at position x and time t. The equation models how heat from a point source diffuses along a thin wire; $k > 0$ represents conductivity of the wire. It can also serve as a model for how opinion distributes over a social media network, which relates to the transition from local data to global data. Applications to opinion dynamics, where a *single* topic is under discussion, first appeared in [50], [72], [142].

3.4.1 Spectral Graph Theory

In Chap. 1, we represented the web as a weighted, directed graph, without giving a formal definition of a graph. We now rectify this:

Definition 3.4.1 A directed graph G is a collection of vertices $V = \{v_i \mid i \in I\}$ and edges $E \subseteq V \times V$. A *loop* is $(v_i, v_i) \in E$, and a multiple edge is an element $(v_i, v_j) \in E$ that appears more than once.

(a) G is finite if V and E are finite.
(b) G is simple if it has no loops or multiple edges.
(c) G is undirected if we introduce a relation on E: $(v_i, v_j) \sim (v_j, v_i)$.

When "graph" is used without an adjective (directed or undirected), the assumption is that G is undirected. All graphs we work with will be finite and simple, with index set $I \subset \mathbb{Z}$.

Definition 3.4.2 Three important matrices associated to a finite simple graph G are the adjacency matrix A_G, the incidence matrix B_G and the Laplacian L_G.

(a) The adjacency matrix A_G is a $|V| \times |V|$ matrix. If G is undirected, then $(A_G)_{ij} = 1 = (A_G)_{ji}$ if $(v_i, v_j) \in E$, and zero otherwise. For a directed graph, we set $(A_G)_{ij} = 1$ if $(v_i, v_j) \in E$, and zero otherwise.
(b) The incidence matrix B_G of G is a $|V| \times |E|$ matrix; columns are indexed by edges and rows by vertices. In a column indexed by edge (v_i, v_j) all entries are zero, except for a -1 in the row indexed by v_i, and a 1 in the row indexed by v_j; (v_i, v_j) represents a directed arrow $v_i \to v_j$.
(c) The Laplacian matrix is defined as $L_G = B_G \cdot B_G^T$.

Example 3.4.3 It will be useful for us to have a running example.

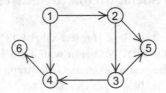

The oriented edges of G are

$$\{(1, 2), (1, 4), (2, 3), (2, 5), (3, 4), (3, 5), (4, 6)\}, \text{ so}$$

$$B_G = \begin{bmatrix} -1 & -1 & 0 & 0 & 0 & 0 & 0 \\ 1 & 0 & -1 & -1 & 0 & 0 & 0 \\ 0 & 0 & 1 & 0 & -1 & -1 & 0 \\ 0 & 1 & 0 & 0 & 1 & 0 & -1 \\ 0 & 0 & 0 & 1 & 0 & 1 & 0 \\ 0 & 0 & 0 & 0 & 0 & 0 & 1 \end{bmatrix} \text{ and } A_G = \begin{bmatrix} 0 & 1 & 0 & 1 & 0 & 0 \\ 0 & 0 & 1 & 0 & 1 & 0 \\ 0 & 0 & 0 & 1 & 1 & 0 \\ 0 & 0 & 0 & 0 & 0 & 1 \\ 0 & 0 & 0 & 0 & 0 & 0 \\ 0 & 0 & 0 & 0 & 0 & 0 \end{bmatrix}$$

The last two rows of A_G are zero because there are no directed edges of G starting at v_5 or v_6. Note that the undirected adjacency matrix for G is just $A_G + A_G^T$. A computation shows that the Laplacian is

$$L_G = \begin{bmatrix} 2 & -1 & 0 & -1 & 0 & 0 \\ -1 & 3 & -1 & 0 & -1 & 0 \\ 0 & -1 & 3 & -1 & -1 & 0 \\ -1 & 0 & -1 & 3 & 0 & -1 \\ 0 & -1 & -1 & 0 & 2 & 0 \\ 0 & 0 & 0 & -1 & 0 & 1 \end{bmatrix}$$

Exercise 3.4.4 The degree $\deg(v)$ of a vertex v is the $\sharp$ of edges incident to v. Prove $(L_G)_{i,j} = \deg(v_i)$ if $i = j$, is -1 if (v_i, v_j) is an edge, and is 0 otherwise. $\diamond$

Lemma 3.4.5 *The Laplacian is symmetric, singular, and positive semidefinite.*

Proof Since $L_G = B_G \cdot B_G^T$, the symmetry of L_G is automatic, and L_G is singular since the column sums are zero. To see that L_G is positive semidefinite, normalize the eigenvectors w_i so that $|w_i| = 1$. For the corresponding eigenvalue λ_i of L_G

$$\begin{aligned} \lambda_i &= w_i \cdot \lambda_i \cdot w_i^T \\ &= w_i \cdot L_G \cdot w_i^T \\ &= (w_i \cdot B_G)(B_G^T \cdot w_i^T) \\ &= (w_i \cdot B_G) \cdot (w_i \cdot B_G)^T \geq 0 \end{aligned}$$

By Exercise 1.4.6, L_G has all real eigenvalues. The set of eigenvalues of a matrix is called the *spectrum*; the spectrum of L_G is $\{0 = \lambda_1 \leq \lambda_2 \leq \cdots \leq \lambda_{|V|}\}$. $\square$

In Chap. 4, you will prove that the number of connected components of G is the dimension of the nullspace of L_G. So if G is connected, $0 < \lambda_2$; this number is called the algebraic connectivity, and is closely related to the smallest *cut* of G. A cut is a partition of V into two disjoint subsets $V_1 \sqcup V_2$, and the cutset is the set of edges having one vertex in V_1 and one vertex in V_2. This turns out to be related to *bottleneck distance*, which appears in Chap. 7.

3.4.2 Heat Diffusing on a Wire Graph

Now we imagine our graph G as a collection of wires, with heat sources placed at the vertices. For a vertex $v_i \in G$ and time t, what is the temperature $u_i(t)$? By Newton's law of cooling, the heat flow along an edge $[v_i, v_j]$ is proportional to the difference in temperature at the two vertices:

$$\frac{du_i}{dt} = -k \cdot \sum_j (A_G)_{ij}(u_i - u_j), \text{ where } k \text{ is the conductivity of the wire.}$$

Now we rearrange things and consolidate

$$
\begin{aligned}
&= &-k \cdot u_i \sum_j (A_G)_{ij} - k \cdot \sum_j (A_G)_{ij}(-u_j) \\
&= &-k \cdot u_i \cdot \deg(v_i) - k \cdot \sum_j (A_G)_{ij}(-u_j) \\
&= &-k \cdot \sum_j (L_G)u_j
\end{aligned}
$$

This yields a system of first order ordinary differential equations, which we can solve by diagonalization. To solve a system of first order ODE's—that is, a system of equations of the form $y_i' = \sum a_i y_i$, with $i \in \{1, \ldots, n\}$—we write the system in the form below, with A an $n \times n$ matrix:

$$\mathbf{y}' = A\mathbf{y}, \text{ where } \mathbf{y} = (y_1, \ldots, y_n)^T, \text{ and } \mathbf{y}' = (y_1', \ldots, y_n')^T.$$

Now let $\mathbf{y} = P\mathbf{w}$ for an $n \times n$ matrix P, so that $\mathbf{y}' = P\mathbf{w}'$, and

$$P\mathbf{w}' = \mathbf{y}' = A\mathbf{y} = AP\mathbf{w} \Rightarrow \mathbf{w}' = P^{-1}AP\mathbf{w}.$$

If we can find a P so that $P^{-1}AP = D$ is diagonal, we reduce to solving a system of equations of the form $w_i' = b_i w_i$, so that $w_i = c_i e^{b_i t}$. So our goal is to diagonalize A, solve for $\mathbf{w}$, then back substitute via $\mathbf{y} = P\mathbf{w}$ to solve for $\mathbf{y}$.

Example 3.4.6 Add the edge $(1, 6)$ to the graph G of Example 3.4.3, and call the new graph H (this is so we have nice eigenvalues: L_G has only 2 integral eigenvalues). If we let $k = 1$, then $-L_H$ has eigenvalues $\{0, -1, -3, -3, -4, -5\}$, so $\mathbf{w} = (c_1, c_2 e^{-t}, c_3 e^{-3t}, c_4 e^{-3t}, c_5 e^{-4t}, c_6 e^{-5t})$, with $\mathbf{y} = P\mathbf{w}$.

$$
-L_H = \begin{bmatrix}
-3 & 1 & 0 & 1 & 0 & 1 \\
1 & -3 & 1 & 0 & 1 & 0 \\
0 & 1 & -3 & 1 & 1 & 0 \\
1 & 0 & 1 & -3 & 0 & 1 \\
0 & 1 & 1 & 0 & -2 & 0 \\
1 & 0 & 0 & 1 & 0 & -2
\end{bmatrix}
\text{ and } P = \begin{bmatrix}
1 & 1 & 1 & 1 & 1 & 1 \\
1 & -1 & 1 & 1 & -1 & -1 \\
1 & -1 & -1 & 0 & -1 & 1 \\
1 & 1 & -1 & 0 & 1 & -1 \\
1 & -2 & 0 & -1 & 1 & 0 \\
1 & 2 & 0 & -1 & -1 & 0
\end{bmatrix}
$$

3.4.3 From Graphs to Cellular Sheaves

How does this relate to sheaves? Since the Laplacian provides a way to understand how a system whose initial data is local (that of the heat sources at vertices) evolves over time, it can also help analyze how opinions on an issue evolve over time. In particular, the opinion at a vertex only impacts the vertices which can "hear" it—those vertices directly adjacent to it. Over time, the opinion diffuses over a larger and larger subset of vertices, just as modeled by the heat equation. If there are opinions on several topics under discussion, then the data structure over a vertex is not just a single number. Hansen–Ghrist introduce *cellular* sheaves as a way to address this.

Definition 3.4.7 ([86]) A cellular sheaf $\mathscr{F}$ on a graph G is

* a vector space $\mathscr{F}(v)$ for each vertex $v \in G$.
* a vector space $\mathscr{F}(e)$ for each edge $e \in G$.
* a linear transformation

$$\mathscr{F}(v) \xrightarrow{\phi_{ev}} \mathscr{F}(e) \text{ if } v \in e.$$

This fits nicely into the machinery of sheaves. In Exercise 3.4.8 you'll show that by choosing $n \gg 0$ it is possible to embed $G \subseteq \mathbb{R}^n$ in such a way that all edges have unit length. Let

$$\mathscr{U} = \{N_{\frac{1}{2}+\epsilon}(v_i) \mid v_i \in V\}$$

and let $\mathscr{U}'$ be the induced open cover of G. Next, we define maps so that there is a complex of vector spaces

$$0 \longrightarrow \bigoplus_{v_i \in V_G} \mathscr{F}(N_{\frac{1}{2}+\epsilon}(v_i)) \longrightarrow \bigoplus_{e_{ij} \in E_G} \mathscr{F}(N_\epsilon(e_{ij})) \longrightarrow 0,$$

or written more succinctly

$$0 \longrightarrow \bigoplus_{v \in V_G} \mathscr{F}(v) \xrightarrow{\delta} \bigoplus_{e \in E_G} \mathscr{F}(e) \longrightarrow 0 \qquad (3.4.2)$$

A first guess might be that the map δ is

$$\delta = \bigoplus_{v \in e} \phi_{ev},$$

but thinking back to the previous section, since sheaves are supposed to capture global consistency, this is not quite right. This is a first example of a Čech complex, which will be defined formally in Sect. 5.1.2. First, orient (if it is not already oriented) G. Then if $e = (v_i, v_j)$, the map which will encode global consistency is exactly the Čech differential:

$$\delta(v_i)|_e = -\phi_{ev_i}$$
$$\delta(v_j)|_e = +\phi_{ev_j}$$

Exercise 3.4.8 Prove the above assertion on embedding G in $\mathbb{R}^n$.   ◇

In Proposition 5.2.3 of Chap. 5 we prove that if

$$V_1 \xrightarrow{d_1} V_2 \xrightarrow{d_2} V_3$$

is a complex of inner product spaces, then

$$V_2 \simeq \text{im}(d_1) \oplus \text{im}(d_2^T) \oplus \text{ker}(L), \text{ where } L = d_1 d_1^T + d_2^T d_2. \tag{3.4.3}$$

In general, the homology of a complex is simply a quotient, which has no canonical basis. The point is that with the extra structure we have in an inner product space, there is a canonical decomposition, and

$$\text{ker}(d_2)/\text{im}(d_1) \simeq \text{ker}(L).$$

Applying this to Eq. 3.4.2, we observe that the global sections of the cellular sheaf $\mathscr{F}$ have a representation as the kernel of the sheaf Laplacian:

$$H^0(\mathscr{F}) = \text{ker}(\delta^T \cdot \delta).$$

This seems very far afield from the nice concrete graph Laplacian we began with, but the exercise below shows that is not the case.

Exercise 3.4.9 Prove that if $\mathscr{F}(v) = \mathscr{F}(e) = \mathbb{K}$, then $L_G = \delta^T \cdot \delta$. Hint: make sure that you make consistent choices for the ordered bases. Next, show that we can measure how far an element $c_0 \in C^0(\mathscr{F})$ is from being in $H^0(\mathscr{F})$: if we let $C^0(\mathscr{F}) \xrightarrow{\pi} H^0(\mathscr{F})$ be orthogonal projection, $d(c_0, H^0(\mathscr{F})) = |c_0 - \pi(c_0)|$.   ◇

Example 3.4.10 Define a sheaf on the graph H of Example 3.4.6 as follows: at each vertex, we have the vector space $V_i = \mathbb{Q}^2$. Vertices represent individuals, and the values at vertex i represent opinions on the following statements:

<blockquote>
The best animal for a pet is a dog
The best weather is cold and snowy
</blockquote>

Suppose -1 represents "strongly disagree" and 1 represents "strongly agree". As time goes by, individuals communicate with adjacent nodes, and opinions change.

After a long period of time, the system will reach equilibrium and the opinion on any issue will be the same at all vertices. What is the status of opinions after a short period of time, for example, before the opinion at vertex 5 and vertex 6 (the most distant vertices) have reached each other? Suppose the opinions on statement 1 are $y(0) = \{0, 0, 0, 0, -1, 1\}$. Note that $w(0) = P^{-1}y(0)$ and

$$P^{-1} = \frac{1}{12}\begin{bmatrix} 2 & 2 & 2 & 2 & 2 & 2 \\ 1 & -1 & -1 & 1 & -2 & 2 \\ 2 & 2 & -4 & -4 & 2 & 2 \\ 2 & 2 & 2 & 2 & -4 & -4 \\ 2 & -2 & -2 & 2 & 2 & -2 \\ 3 & -3 & 3 & -3 & 0 & 0 \end{bmatrix}.$$

Therefore

$$(c_1, \cdots, c_6) = w(0) = P^{-1}y(0) = \left(0, \frac{1}{3}, 0, 0, -\frac{1}{3}, 0\right),$$

hence $w(t) = (0, \frac{1}{3}e^{-t}, 0, 0, -\frac{1}{3}e^{-4t}, 0)$, and

$$y(t) = Pw(t) = \frac{1}{3}\begin{bmatrix} e^{-t} - e^{-4t} \\ -e^{-t} + e^{-4t} \\ -e^{-t} + e^{-4t} \\ e^{-t} - e^{-4t} \\ -2e^{-t} - e^{-4t} \\ 2e^{-t} + e^{-4t} \end{bmatrix}.$$

Exercise 3.4.11 Compute $y(t)$ for $t \in \{0, 1, 2, 3, \infty\}$ for Example 3.4.10, and explain why the values make sense. We only computed the values for statement 1. For statement 2, use initial data $(0, 0, 0, 0, 1, -1)$ and compute. Carry out the same construction with other sets of initial data.

There is (a priori) no connection between opinions on statement 1 and statement 2. Find some choices for data where there seems to be a relationship between the opinions (e.g. people who like dogs also like cold, snowy weather), and compute the correlation matrix as in Chap. 1 to see if your observation has validity. How do opinions evolve over time?   ◇

Chapter 4
Homology I: Simplicial Complexes to Sensor Networks

This chapter is the fulcrum of the book, the beginning of a beautiful friendship between algebra and topology. The idea is the following: small open neighborhoods of a point on the circle S^1 and a point on the line $\mathbb{R}^1$ are homeomorphic—an ant living in either neighborhood would be unable to distinguish between them. But from a global perspective we can easily spot the difference between $\mathbb{R}^1$ and S^1. Homology provides a way to quantify and compute the difference.

The first step is to model a topological space combinatorially. We do this with *simplicial complexes*, which are a lego set for topologists. The second step is to translate the combinatorial data of a simplicial complex into an *algebraic* complex, as in Eq. 2.2.1 of Chap. 2. Simplicial homology is the homology of the resulting complex; we also discuss singular homology, which is based on formal combinations of maps from a simplex into a topological space.

In Sect. 4.3 we introduce the basics of homological algebra, and in Sect. 4.4 we connect back to topology via the Mayer–Vietoris sequence, which describes how building a simplicial complex from constituent parts translates into algebra, and introduce Rips–Vietoris and Čech complexes, which play a central role in data analysis. The topics covered in this chapter are:

- Simplicial Complexes, Nerve of a Cover.
- Simplicial and Singular Homology.
- Snake Lemma and Long Exact Sequence in Homology.
- Mayer–Vietoris, Rips–Vietoris and Čech complexes, Sensor Networks.

H. Schenck, *Algebraic Foundations for Applied Topology and Data Analysis*, Mathematics of Data 1, https://doi.org/10.1007/978-3-031-06664-1_4
63

4.1 Simplicial Complexes, Nerve of a Cover

Simplices are natural candidates for building blocks in topology:

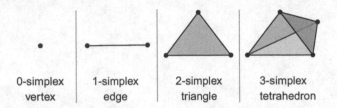

| 0-simplex | 1-simplex | 2-simplex | 3-simplex |
| vertex | edge | triangle | tetrahedron |

Simplices may be thought of as combinatorial or geometric objects:

Definition 4.1.1 An n-simplex σ_n is an $n + 1$ set and all subsets. A *geometric realization* of an n-simplex is the convex hull (set of all convex combinations) C of $n + 1$ points in $\mathbb{R}^d$, such that C is n-dimensional (so $d \geq n$). The *standard* n-simplex is the convex hull of the coordinate points in $\mathbb{R}^{n+1}$.

A *simplicial complex* Δ is constructed by attaching a group of simplices to each other, with the caveat that the attaching is done by identifying shared sub-simplices; sub-simplices are often called *faces*.

Definition 4.1.2 An abstract simplicial complex Δ on a finite vertex set V is a collection of subsets of V such that

- $v \in \Delta$ if $v \in V$.
- $\tau \subseteq \sigma \in \Delta \Rightarrow \tau \in \Delta$.

Example 4.1.3 The second condition above says that to specify a simplicial complex, it suffices to describe the maximal faces. For example, we visualize the simplicial complex on $V = \{0, 1, 2, 3\}$ with maximal faces denoted $[0, 1, 2]$ and $[1, 3]$, corresponding to a triangle with vertices labelled $\{0, 1, 2\}$ and an edge with vertices labelled $\{1, 3\}$. For a non-example, consider the same triangle, but where the vertex labelled 1 from edge $\{1, 3\}$ is glued to the middle of the edge $\{1, 2\}$.

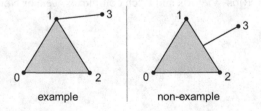

example   non-example

Definition 4.1.4 The dimension of Δ is the dimension of the largest face of Δ. An *orientation* on Δ is a choice of ordering of the vertices of σ for each $\sigma \in \Delta$. Oriented one and two simplices are depicted below.

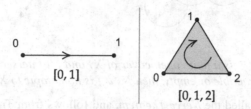

We will use square brackets to denote oriented simplices.

4.1.1 The Nerve of a Cover

A key construction in topology connects a cover to a simplicial complex. Recall from Definition 3.1.7 that for a topological space X, an open cover of X is a collection of open sets satisfying

$$\mathcal{U} = \{U_i \mid i \in I\}, \text{ with } \bigcup_{i \in I} U_i = X$$

Definition 4.1.5 For an open cover $\mathcal{U}$ of a space X with index set I, let I_k denote the set of all $k+1$-tuples of I. The *nerve* $N(\mathcal{U})$ of $\mathcal{U}$ is a simplicial complex whose k-faces $N(\mathcal{U})_k$ are given by

$$N(\mathcal{U})_k \iff \{\alpha_k \in I_k \mid (\bigcap_{i \in \alpha_k} U_i) \neq \emptyset\}$$

To see that $N(\mathcal{U})$ is a simplicial complex, note that since if $\alpha_k \in I_k$ and β_{k-1} is a k-subset of α_k, this implies that

$$(\bigcap_{i \in \beta_{k-1}} U_i) \neq \emptyset$$

Example 4.1.6 For the four open sets of the cover below, $U_1 \cap U_2 \cap U_3 \neq \emptyset$, yielding the 2-simplex [123], while U_4 only meets U_3, yielding the edge [34].

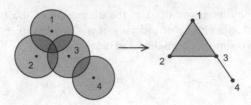

Theorem 4.1.7 *If $\mathscr{U}$ is a finite open cover of X, and all intersections of the open sets of $\mathscr{U}$ are contractible or empty, then $N(\mathscr{U})$ is homotopic to X.*

Theorem 4.1.7 is called the *Nerve Theorem*, and follows from Theorem 9.4.5; see also Exercise 5.1.5. In Sect. 4.4 of this chapter, we'll see two standard ways to build a simplicial complex Δ_ϵ from the space X_ϵ appearing in the introduction; the first way is via the nerve of a cover. The Nerve Theorem also holds for a closed cover of a CW-complex (which are defined in Sect. 5.3) by subcomplexes, see [16].

4.2   Simplicial and Singular Homology

We recall the definition of an *algebraic* complex from Chap. 2: a complex $\mathscr{C}$ is a sequence of modules and homomorphisms

$$\mathscr{C}_\bullet : \cdots M_{j+1} \xrightarrow{d_{j+1}} M_j \xrightarrow{d_j} M_{j-1} \cdots$$

such that $\operatorname{im}(d_{j+1}) \subseteq \ker(d_j)$; the jth homology is $H_j(\mathscr{C}) = \ker(d_j)/\operatorname{im}(d_{j+1})$.

Definition 4.2.1 Let Δ be an oriented simplicial complex, and let R be a ring. Define $C_i(\Delta, R)$ as the free R-module with basis the oriented i-simplices, modulo the relations

$$[v_0, \ldots, v_n] \sim (-1)^{\operatorname{sgn}(\sigma)}[v_{\sigma(0)}, \ldots, v_{\sigma(n)}],$$

where $\sigma \in S_{n+1}$ is a permutation. Any permutation σ can be written as a product of transpositions (two cycles); the sign of σ is one if the number of transpositions is odd, and zero if the number of transpositions is even.

Example 4.2.2 For an edge, this says

$$[v_0, v_1] = -[v_1, v_0]$$

This is familiar from vector calculus: when it comes to integration

$$\int_{v_0}^{v_1} f(x)dx = -\int_{v_1}^{v_0} f(x)dx$$

For a triangle, the orientation $[v_2, v_1, v_0]$ requires 3 transpositions to change to $[v_0, v_1, v_2]$, hence

$$[v_2, v_1, v_0] = -[v_0, v_1, v_2]$$

Definition 4.2.3 Let ∂ be the map from $C_i(\Delta, R)$ to $C_{i-1}(\Delta, R)$ defined by

$$\partial([v_{j_0}, \dots, v_{j_i}]) = \sum_{k=0}^{i} (-1)^k [v_{j_0}, \dots, \widehat{v_{j_k}}, \dots, v_{j_i}], \text{ where } \widehat{v} \text{ means omit } v.$$

The map ∂ is called the *boundary* map; it sends an i-simplex to a signed sum of its $i - 1$ faces. Applying ∂^2 to a basis element shows $\partial^2 = 0$, so the C_i form a chain complex $\mathscr{C}$. Simplicial homology $H_i(\Delta, R)$ is the homology of $\mathscr{C}$.

Example 4.2.4 If R is a field, then the C_i form a complex of vector spaces V_i.

$$\mathscr{V}: 0 \longrightarrow V_n \longrightarrow V_{n-1} \longrightarrow \cdots \longrightarrow V_0 \longrightarrow 0.$$

An induction shows that

$$\sum_{i=0}^{n} (-1)^i \dim V_i = \sum_{i=0}^{n} (-1)^i \dim H_i(\mathscr{V}).$$

The alternating sum $\chi(\mathscr{V})$ above is known as the *Euler characteristic*.

Example 4.2.5 We return to our pair of ants, one living on S^1 and one on $\mathbb{R}^1$. Let $\mathbb{K}$ be a field, and model S^1 and (a portion of) $\mathbb{R}^1$ with oriented simplicial complexes Δ_1 and Δ_2 having maximal faces

$$\Delta_1 = \{[0, 1], [1, 2], [2, 0]\} \text{ and } \Delta_2 = \{[0, 1], [1, 2], [2, 3]\}$$

Let $C_1(\Delta_i, \mathbb{K})$ have the ordered bases above, and order the basis of $C_0(\Delta_1, \mathbb{K})$ as $\{[0], [1], [2]\}$. With respect to these bases, the only nonzero differential is ∂_1:

$$\partial_1 = \begin{bmatrix} -1 & 0 & 1 \\ 1 & -1 & 0 \\ 0 & 1 & -1 \end{bmatrix}$$

$$C_\bullet(\Delta_1, \mathbb{K}) = \quad 0 \longrightarrow \mathbb{K}^3 \longrightarrow \mathbb{K}^3 \longrightarrow 0$$

and we compute that $H_1(\Delta_1) = \ker(\partial_1) = [1, 1, 1]^T$. On the other hand, for Δ_2, using $\{[0], [1], [2], [3]\}$ as the ordered basis for $C_0(\Delta_2)$, the complex is

$$\partial_1 = \begin{bmatrix} -1 & 0 & 0 \\ 1 & -1 & 0 \\ 0 & 1 & -1 \\ 0 & 0 & 1 \end{bmatrix}$$

$$C_\bullet(\Delta_2, \mathbb{K}) = \quad 0 \longrightarrow \mathbb{K}^3 \xrightarrow{\hspace{4cm}} \mathbb{K}^4 \longrightarrow 0$$

and we compute that $H_1(\Delta_2) = 0$. So H_1 distinguishes between Δ_1 and Δ_2.

Exercise 4.2.6 The example above illustrates *the* key point about homology: For $k \geq 1$, H_k measures k-dimensional *holes*. Draw a picture and compute homology when Δ has oriented simplices $\{[0, 1], [1, 2], [2, 0], [2, 3], [0, 3]\}$ ◇

Exercise 4.2.7 If we augment $\mathscr{C}$ by adding in the empty face so $C_{-1}(\Delta, R) = R$, the result is *reduced homology* $\widetilde{H}_i(\Delta, R)$. Show $\widetilde{H}_i(\Delta, R) = H_i(\Delta, R)$ for $i \geq 1$, and $\widetilde{H}_0(\Delta, R) = 0$ iff Δ is connected.   ◇

Simplicial homology is a topological invariant, which can be proved by taking common refinements of two different triangulations. For an infinite simplicial complex, we define elements of $C_i(\Delta, R)$ as functions c from the i-simplices to R which are zero except on finitely many simplices, such that if τ and τ' are the same as *unoriented* simplices and $\tau = \sigma(\tau')$ then $c(\tau) = \mathrm{sgn}(\sigma)c(\tau')$. This will be useful in Sect. 4.2.1, where we define *singular* homology, for which topological invariance is automatic. The downside is that singular homology is difficult to compute.

Example 4.2.8 We compute the simplicial homology of the sphere S^2, using two different triangulations. Let Δ_1 be the simplicial complex defined by the boundary of a tetrahedron, having maximal faces $\{[1, 2, 3], [1, 2, 4], [1, 3, 4], [2, 3, 4]\}$, and let Δ_2 be the simplicial complex consisting of two hollow tetrahedra, glued along a common triangular face, with the glued face subsequently removed; Δ_2 has maximal faces $\{[1, 2, 4], [1, 3, 4], [2, 3, 4], [1, 2, 5], [1, 3, 5], [2, 3, 5]\}$.

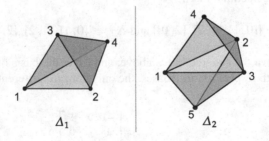

With oriented, ordered basis for $C_2(\Delta_1, R)$ as above, oriented basis for $C_1(\Delta_1, R) = \{[1, 2], [1, 3], [1, 4], [2, 3], [2, 4], [3, 4]\}$, and basis for $C_0(\Delta_1, R) = \{[1], [2], [3], [4]\}$, the complex $C_\bullet(\Delta_1, R)$ is given by

$$0 \xrightarrow{} R^4 \xrightarrow{\partial_2} R^6 \xrightarrow{\partial_1} R^4 \xrightarrow{} 0$$

with

$$\partial_2 = \begin{bmatrix} 1 & 1 & 0 & 0 \\ -1 & 0 & 1 & 0 \\ 0 & -1 & -1 & 0 \\ 1 & 0 & 0 & 1 \\ 0 & 1 & 0 & -1 \\ 0 & 0 & 1 & 1 \end{bmatrix} \text{ and } \ker(\partial_2) = [1, -1, 1, -1]^T$$

and

$$\partial_1 = \begin{bmatrix} -1 & -1 & -1 & 0 & 0 & 0 \\ 1 & 0 & 0 & -1 & -1 & 0 \\ 0 & 1 & 0 & 1 & 0 & -1 \\ 0 & 0 & 1 & 0 & 1 & 1 \end{bmatrix}$$

We compute $\ker(\partial_1) = \mathrm{im}(\partial_2)$, and $\mathrm{im}(\partial_1) \simeq R^3 \subseteq R^4 = \ker(\partial_0)$, hence

$$\begin{aligned} H_2(\Delta_1, R) &\simeq R^1 \\ H_1(\Delta_1, R) &\simeq 0 \\ H_0(\Delta_1, R) &\simeq R^1 \end{aligned}$$

Exercise 4.2.9 Compute $H_i(\Delta_2, R)$. You should obtain the same values for homology as above: $H_1(\Delta_2, R)$ vanishes, and $H_2(\Delta_2, R) \simeq R \simeq H_0(\Delta_2, R)$. We'll see in Chap. 5 that this symmetry is a consequence of Poincaré duality. $\diamond$

The simplest possible model for S^2 would consist of two triangles, glued together along the 3 common edges. The definition of CW complexes in Chap. 5 will allow this, and in fact permit an even simpler model for S^2.

4.2.1 Singular homology

In singular homology, the objects of study are maps to a topological space X. A *singular n-simplex* is a continuous map from the standard n-simplex Δ_n to X. The free abelian group on the singular n-simplices on X is the *singular n-chains* $S_n(X)$.

If $\sigma \in S_n(X)$, then restricting σ to an $n-1$ face of Δ_n yields (after an appropriate change of coordinates) an element of $S_{n-1}(X)$, and we can build a chain complex just as we did with simplicial homology:

Definition 4.2.10 Let r_i be the map that changes coordinates from the basis vectors $\{e_1, \ldots, \widehat{e_i}, \ldots, e_{n+1}\}$ to the basis vectors $\{e_1, \ldots, e_n\}$. For

$$f \in S_n(X), \text{ let } f_i \text{ be the restriction of } f \text{ to the } i^{th} \text{ face of } \Delta_n,$$

and define a complex $\mathscr{S} : \cdots \longrightarrow S_n(X) \xrightarrow{\partial} S_{n-1}(X) \longrightarrow \cdots$ via

$$\partial(f) = \sum_{i=1}^{n+1} (-1)^i r_i f_i.$$

The r_i are needed to shift the domain of definition for the restricted map f_i to the standard $n-1$ simplex, which is the source for the $n-1$ chains. A check shows that $\partial^2 = 0$, and singular homology is defined as the homology of the complex $\mathscr{S}$.

Theorem 4.2.11 *If X and Y are homotopic, then $H_i(X) \simeq H_i(Y)$.*

Proof (**Sketch**) If f is map between topological spaces or simplicial complexes, it gives rise to an induced map f_* on homology. A pair of maps f_*, g_* between *chain complexes* are homotopic if Definition 4.3.7 holds. Theorem 4.3.8 shows homotopic maps of chain complexes induce the same map on homology. The result follows by Exercise 7.4.6, which connects topological & algebraic homotopy.   □

We noted earlier that it is not obvious that simplicial homology is a topological invariant. In fact, there is an isomorphism between the singular and simplicial homology of a space X, and we close with a sketch of the proof.

Theorem 4.2.12 *For a topological space X which can be triangulated, there is an isomorphism from the simplicial homology of X to the singular homology of X.*

Proof (**Sketch**) Let K be a triangulation of X, so $S_i(X) = S_i(K)$. Then the map

$$\phi : C_i(K, R) \longrightarrow S_i(X)$$

defined by sending an i-simplex $\sigma \in K_i$ to the map taking the standard i-simplex to σ commutes with the differentials of $C_\bullet$ and $S_\bullet$, so by Lemma 4.3.4 of this chapter there is an induced map

$$\phi_* : H_i(K, R) \longrightarrow H_i(X)$$

To show the map is an isomorphism, first prove it when K is a finite simplicial complex, by inducting on the number of simplices. When K is infinite, a class in singular homology is supported on a compact subset C. Then one shows that there is a finite subcomplex of K containing C, reducing back to the finite case. □

4.3 Snake Lemma and Long Exact Sequence in Homology

This section lays the groundwork for the rest of the chapter. The centerpiece is the existence of a long exact sequence in homology arising from a short exact sequence of complexes, and the key ingredient in proving this is the *Snake Lemma*.

4.3.1 Maps of complexes, Snake Lemma

A diagram is *commutative* if the maps commute. Informally, this means that going "first right then down" yields the same result as going "first down, then right".

Lemma 4.3.1 (The Snake Lemma) *For a commutative diagram of R–modules with exact rows*

$$
\begin{array}{ccccccc}
A_1 & \xrightarrow{a_1} & A_2 & \xrightarrow{a_2} & A_3 & \longrightarrow & 0 \\
\downarrow{f_1} & & \downarrow{f_2} & & \downarrow{f_3} & & \\
0 \longrightarrow & B_1 & \xrightarrow{b_1} & B_2 & \xrightarrow{b_2} & B_3 &
\end{array}
$$

there is an exact sequence:

$$
\begin{array}{ccc}
\ker f_1 \longrightarrow \ker f_2 \longrightarrow \ker f_3 \\
\swarrow{\delta} \\
\operatorname{coker} f_1 \longrightarrow \operatorname{coker} f_2 \longrightarrow \operatorname{coker} f_3
\end{array}
$$

The name comes from the connecting homomorphism δ, which *snakes* from the end of the top row to the start of the bottom row.

Proof The key to the snake lemma is defining the connecting homomorphism δ : $\ker(f_3) \to \mathrm{coker}(f_1)$. Let $a \in \ker(f_3)$. Since the map a_2 is surjective, there exists $b \in A_2$ such that $a_2(b) = a$, hence

$$f_3(a_2(b)) = f_3(a) = 0.$$

Since the diagram commutes,

$$0 = f_3(a_2(b)) = b_2(f_2(b)).$$

By exactness of the rows, $f_2(b) \in \ker(b_2) = \mathrm{im}(b_1)$. So there exists $c \in B_1$ with

$$b_1(c) = f_2(b).$$

Let $\tilde{c}$ be the image of c in $\mathrm{coker}(f_1) = B_1/\mathrm{im}(f_1)$, and define

$$\delta(a) = \tilde{c}.$$

To show that δ is well defined, consider where choices were made: if $a_2(b') = a$, then $a_2(b - b') = 0$, hence $b - b' \in \ker(a_2) = \mathrm{im}(a_1)$, so there exists $e \in A_1$ with $a_1(e) = b - b'$, hence $b = b' + a_1(e)$. Applying f_2 we have $f_2(b) = f_2(b') + f_2(a_1(e)) = f_2(b') + b_1(f_1(e))$. But b_1 is an inclusion, so $f_2(b)$ and $f_2(b')$ agree modulo $\mathrm{im}(f_1)$, so δ is indeed a well defined map to $\mathrm{coker}(f_1)$.   □

Exercise 4.3.2 Check that the Snake is exact.   ◇

Definition 4.3.3 If $A_\bullet$ and $B_\bullet$ are complexes, then a ***morphism of complexes*** ϕ is a family of homomorphisms $A_i \overset{\phi_i}{\to} B_i$ making the diagram below commute:

$$
\begin{array}{ccccccccc}
A_\bullet: & \cdots \longrightarrow & A_{i+1} & \overset{\partial_{i+1}}{\longrightarrow} & A_i & \overset{\partial_i}{\longrightarrow} & A_{i-1} & \overset{\partial_{i-1}}{\longrightarrow} & \cdots \\
 & & \downarrow \phi_{i+1} & & \downarrow \phi_i & & \downarrow \phi_{i-1} & & \\
B_\bullet: & \cdots \longrightarrow & B_{i+1} & \underset{\delta_{i+1}}{\longrightarrow} & B_i & \underset{\delta_i}{\longrightarrow} & B_{i-1} & \underset{\delta_{i-1}}{\longrightarrow} & \cdots
\end{array}
$$

Lemma 4.3.4 (Induced Map on Homology) *A morphism of complexes induces a map on homology.*

Proof To show that ϕ_i induces a map $H_i(A_\bullet) \to H_i(B_\bullet)$, take $a_i \in A_i$ with $\partial_i(a_i) = 0$. Since the diagram commutes,

$$0 = \phi_{i-1}\partial_i(a_i) = \delta_i\phi_i(a_i)$$

Hence, $\phi_i(a_i)$ is in the kernel of δ_i, so we obtain a map $\ker \partial_i \to H_i(B_\bullet)$. If $a_i = \partial_{i+1}(a_{i+1})$, then

$$\phi_i(a_i) = \phi_i\partial_{i+1}(a_{i+1}) = \delta_{i+1}\phi_{i+1}(a_{i+1}),$$

so ϕ takes the image of ∂ to the image of δ, yielding a map $H_i(A_\bullet) \to H_i(B_\bullet)$. $\square$

Definition 4.3.5 A *short exact sequence of complexes* is a commuting diagram:

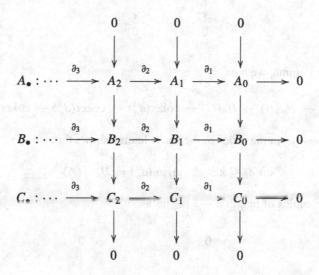

where the columns are exact and the rows are complexes.

We now come to the fundamental result of this section.

Theorem 4.3.6 (Long Exact Sequence in Homology) *A short exact sequence of complexes yields a long exact sequence in homology:*

$$\cdots \longrightarrow H_{n+1}(C) \longrightarrow H_n(A) \longrightarrow H_n(B) \longrightarrow H_n(C) \longrightarrow H_{n-1}(A) \longrightarrow \cdots$$

Proof We induct on the length of the complexes. When the diagram above has only two columns, we are in the setting of the Snake Lemma, yielding the base case. So now suppose the result holds for complexes of length $n - 1$, and consider a complex of length n. Prune the two leftmost columns off the short exact sequence of complexes in Definition 4.3.5

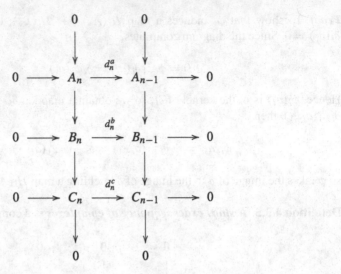

By the Snake Lemma, we have

$$0 \to H_n(A) \to H_n(B) \to H_n(C) \xrightarrow{\delta} \operatorname{coker}(d_n^a) \to \operatorname{coker}(d_n^b) \to \operatorname{coker}(d_n^c) \to 0$$

A diagram chase as in the proof of the snake lemma shows that

$$\operatorname{im}(\delta) \subseteq \ker(d_{n-1}^a)/\operatorname{im}(d_n^a) = H_{n-1}(A).$$

The pruned complex of length $n-1$ begins as

By the Induction Hypothesis, we have a long exact sequence

$$0 \longrightarrow H_{n-1}(A)/\mathrm{im}(H_n(C)) \longrightarrow H_{n-1}(B) \longrightarrow H_{n-1}(C) \longrightarrow \cdots \qquad (4.3.1)$$

Since $\mathrm{im}(\delta) \subseteq H_{n-1}(A)$, we also have an exact sequence

$$\cdots \rightarrow H_n(C) \xrightarrow{\delta} H_{n-1}(A) \rightarrow H_{n-1}(A)/H_n(C) \rightarrow 0. \qquad (4.3.2)$$

To conclude, splice the surjection of Eq. 4.3.2

$$H_{n-1}(A) \rightarrow H_{n-1}(A)/H_n(C) \rightarrow 0$$

with the inclusion of Eq. 4.3.1

$$0 \rightarrow H_{n-1}(A)/H_n(C) \rightarrow H_{n-1}(B),$$

to obtain a map $H_{n-1}(A) \rightarrow H_{n-1}(B)$, which completes the induction. □

4.3.2 Chain Homotopy

When do two morphisms of complexes induce the same map on homology?

Definition 4.3.7 If A and B are complexes, and α, β are morphisms of complexes, then α and β are **homotopic** if there exists a family of homomorphisms $A_i \xrightarrow{\gamma_i} B_{i+1}$ such that for all i, $\alpha_i - \beta_i = \delta_{i+1}\gamma_i + \gamma_{i-1}\partial_i$; a diagram illustrating this appears in Exercise 7.4.6. Notice that γ need not commute with ∂ and δ.

Theorem 4.3.8 *Homotopic maps induce the same map on homology.*

Proof It suffices to show that if $\alpha_i = \delta_{i+1}\gamma_i + \gamma_{i-1}\partial_i$ then α induces the zero map on homology. But if $a_i \in H_i(A)$, then since $\partial_i(a_i) = 0$,

$$\alpha_i(a_i) = \delta_{i+1}\gamma_i(a_i) + \gamma_{i-1}\partial_i(a_i) = \delta_{i+1}\gamma_i(a_i) \in \mathrm{im}(\delta),$$

so $\alpha_i(a_i) = 0$ in $H_i(B)$. □

4.4 Mayer–Vietoris, Rips and Čech Complex, Sensor Networks

As we noted at the beginning of the chapter, homological algebra grew out of topology, and we illustrate this now.

4.4.1 Mayer–Vietoris Sequence

A central idea in mathematics is to study an object by splitting it into simpler component parts. The Mayer–Vietoris sequence provides a way to do this in the topological setting. Basically, Mayer–Vietoris is a "divide and conquer" strategy. Given a topological space or simplicial complex Z, if we write

$$Z = X \cup Y$$

then there will be relations between X, Y, Z and the intersection $X \cap Y$.

Theorem 4.4.1 (Mayer–Vietoris) *Let X and Y be simplicial complexes, and let $Z = X \cup Y$. Then there is a short exact sequence of complexes*

$$0 \longrightarrow C_\bullet(X \cap Y) \longrightarrow C_\bullet(X) \oplus C_\bullet(Y) \longrightarrow C_\bullet(X \cup Y) \longrightarrow 0,$$

yielding a long exact sequence in homology.

Proof We need to show that for a fixed index i, there are short exact sequences

$$0 \longrightarrow C_i(X \cap Y) \longrightarrow C_i(X) \oplus C_i(Y) \longrightarrow C_i(X \cup Y) \longrightarrow 0$$

Define the map from

$$C_i(X) \oplus C_i(Y) \longrightarrow C_i(X \cup Y) \text{ via } (\sigma_1, \sigma_2) \mapsto \sigma_1 - \sigma_2$$

The kernel of the map is exactly $C_i(X \cap Y)$.   □

Example 4.4.2 In Example 4.2.8, the complex Δ_2 is constructed from a pair of hollow tetrahedra by dropping a triangle from each one, then identifying the two "deleted" tetrahedra along the edges of the deleted triangle. From a topological perspective, this is equivalent to gluing two disks along their boundaries. The top dimensional faces of the complexes involved are

$$
\begin{aligned}
X &= & \{[1,2,4], [1,3,4], [2,3,4]\} \\
Y &= & \{[1,2,5], [1,3,5], [2,3,5]\} \\
X \cap Y &= & \{[1,2], [1,3], [2,3]\} \\
X \cup Y &= & \{[1,2,4], [1,3,4], [2,3,4], [1,2,5], [1,3,5], [2,3,5]\}
\end{aligned}
$$

So our short exact sequence of complexes, with R coefficients, has the form

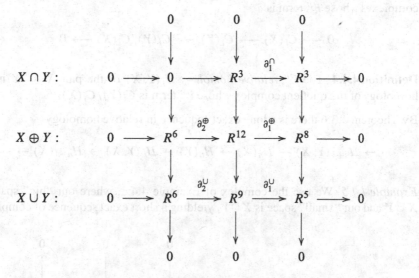

The ∂_i are written with superscripts to distinguish them. With oriented basis for $C_1(X \cap Y)$ as above and basis for $C_0(X \cap Y) = \{[1], [2], [3]\}$ we have

$$\partial_1^{\cap} = \begin{bmatrix} -1 & -1 & 0 \\ 1 & 0 & -1 \\ 0 & 1 & 1 \end{bmatrix}$$

So we have

$$H_1(X \cap Y) \simeq R^1 \simeq H_0(X \cap Y).$$

An easy check shows that for both X and Y, $H_i = 0$ if $i \neq 0$, so from the long exact sequence in homology we have

$$H_2(X \cup Y) \simeq H_1(X \cap Y).$$

Exercise 4.4.3 Compute the differentials above and the induced maps between the rows. Check that the diagram commutes, and calculate the homology directly. ◇

4.4.2 Relative Homology

When $X \subseteq Y$, a basic topological question is how the homology of X and Y are related. At the level of simplicial complexes,

$$C_i(X) \subseteq C_i(Y)$$

so the most straightforward approach is to consider the short exact sequence of complexes whose i^{th} term is

$$0 \longrightarrow C_i(X) \longrightarrow C_i(Y) \longrightarrow C_i(Y)/C_i(X) \longrightarrow 0$$

Definition 4.4.4 The ith *relative homology* $H_i(Y, X)$ of the pair $X \subseteq Y$ is the homology of the quotient complex whose i^{th} term is $C_i(Y)/C_i(X)$.

By Theorem 4.3.6 there is a long exact sequence in relative homology

$$\cdots \to H_{n+1}(Y, X) \to H_n(X) \to H_n(Y) \to H_n(Y, X) \to H_{n-1}(X) \to \cdots$$

Example 4.4.5 We use the complex of Example 4.4.2, where our "big" space is $X \cup Y$ and our "small" space is $X \cap Y$, yielding a short exact sequence of complexes

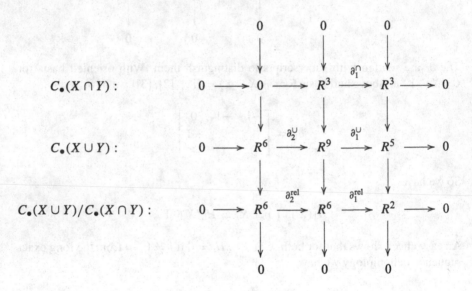

The horizontal differentials in the top two rows are the same as the differentials $\partial_i^\cap$ and $\partial_i^\cup$ in Example 4.4.2, and the vertical map from the first row to the second row is the inclusion map. Example 4.4.2 showed that $H_2(X \cup Y) \simeq R^1$ and in Exercise 4.4.3, we calculated that $H_1(X \cup Y) = 0$. The long exact sequence yields

$$0 \to H_2(X \cup Y) \to H_2(Y, X) \to H_1(X \cap Y) \to H_1(X \cup Y) = 0.$$

Therefore if R is a field, $H_2(Y, X) \simeq R^2$. In relative homology, the "belt" $X \cap Y \sim S^1$ around the "belly" of $X \cup Y \sim S^2$ is tightened down to a point p, yielding a pair of S^2's (which touch at p).

Exercise 4.4.6 Write out the differentials ∂_i^{rel} and the vertical maps, check everything commutes, and compute homology.   ◇

4.4.3   Čech Complex and Rips Complex

Topological data analysis starts with point cloud data; as outlined in the preface, we introduce an ϵ-neighborhood around each point. As ϵ increases, we obtain a family of topological spaces X_ϵ.

We now take the first step towards turning the spaces X_ϵ into a related family of simplicial complexes Δ_ϵ: for each space X_ϵ, we build a corresponding simplicial complex Δ_ϵ. In the setting of persistent homology, there are two standard ways to build Δ_ϵ; the first method is via the nerve of a cover.

Definition 4.4.7 The *Čech complex* $\mathscr{C}_\epsilon$ has k-simplices which correspond to $k + 1$-tuples of points $p \in X$ such that

$$\bigcap_{i=0}^{k} \overline{N_{\frac{\epsilon}{2}}(p_i)} \neq \emptyset.$$

The $k + 1$ closed balls of radius $\frac{\epsilon}{2}$ centered at $\{p_0, \ldots, p_k\}$ share a common point.

It follows from Theorem 4.1.7 that there is a homotopy $\mathscr{C}_\epsilon \simeq X_{\frac{\epsilon}{2}}$.

Definition 4.4.8 The *Rips (or Rips–Vietoris) complex* $\mathscr{R}_\epsilon$ has k-simplices which correspond to $k + 1$-tuples of points $p \in X$ such that $d(p_i, p_j) \leq \epsilon$ for all pairs i, j.

For a simple graph Γ, the *flag complex* Δ_Γ is a simplicial complex with an m-face for every K_{m+1} subgraph of Γ. The Rips complex is an example of a flag complex.

The Rips complex is often more manageable from a computational standpoint, while the Čech complex often makes proofs simpler. In [53], de Silva–Ghrist show that there is a nice relationship between the two:

Theorem 4.4.9 *There are inclusions* $\mathscr{R}_\epsilon \hookrightarrow \mathscr{C}_{\sqrt{2}\epsilon} \hookrightarrow \mathscr{R}_{\sqrt{2}\epsilon}$.

Example 4.4.10 We compute the Čech and Rips complexes for X consisting of the four points $\{(0, 0), (1, 0), (0, 1), (1, 1)\}$. For $\epsilon < 1$ both $\mathscr{C}_\epsilon$ and $\mathscr{R}_\epsilon$ consist only of the points themselves. For $\epsilon = 1$, we have that

$$\bigcup_{i=1}^{4} \overline{N_{\frac{1}{2}}(p_i)}$$

is the union of four closed balls, tangent at four points

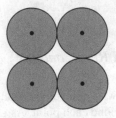

Hence $\mathscr{C}_1$ and $\mathscr{R}_1$ are both (hollow) squares, as below

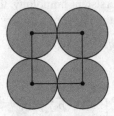

They remain hollow squares until $\epsilon = \sqrt{2}$, when all four balls meet at a common point. Hence $\mathscr{C}_{\sqrt{2}}$ and $\mathscr{R}_{\sqrt{2}}$ are solid tetrahedra.

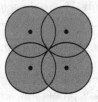

On the other hand, if X consists of the vertices of an equilateral triangle with sides of length one, $\mathscr{C}_{\frac{1}{2}}$ is a hollow triangle, whereas $\mathscr{R}_{\frac{1}{2}}$ is a solid triangle.

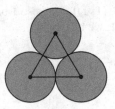

Exercise 4.4.11 Let X_n consist of vertices $\{p_1, \ldots, p_n\}$ of a symmetric n-gon, with $d(0, p_i) = 1$. Carry out the computation of Example 4.4.10 for X_n. ◇

We close with an application of de Silva–Ghrist [52] of homological algebra to the study of sensor networks.

Example 4.4.12 Suppose $D \subset \mathbb{R}^2$ is a fenced in region in the plane (hence compact and connected), with the fence ∂D consisting of a finite number of line segments:

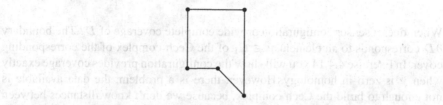

The region is called D because it contains a trove of diamonds. To prevent theft, the owner has deployed a sensor network. As a first step, at each vertex on ∂D, there is a fixed sensor; so we have coverage as depicted in the following diagram:

Within the region D, there is a swarm of wandering sensors s_i; let $\mathscr{U} = \{N_{r_b}(s_i)\}$ and U be the union. The sensors have the following properties

(a) Each sensor transmits a unique ID number, and can detect the identity of another node within transmit distance r_b.
(b) Each sensor has coverage (detects intruders) in a disk centered at the sensor, of radius $r_c = \frac{r_b}{\sqrt{3}}$ (more on this below).
(c) Each of the fixed sensors located at a vertex of ∂D knows the ID of its neighbors, which lie within radius r_b.

For certain configurations of the sensors, there will be coverage of all of D. We add 3 roving sensors to the example above. In the right hand figure below, the resulting coverage is complete. But in the left hand figure, there is a gap in the coverage, denoted by the darkened zone in the southeast region.

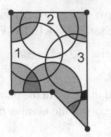 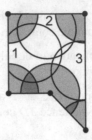

When does a sensor configuration provide complete coverage of D? The boundary ∂D corresponds to an element $\gamma \in C_1$ of the Čech complex of the corresponding cover. In Exercise 4.4.14 you will show the configuration provides coverage exactly when γ is zero in homology. However, there is a problem: the data available is not enough to build the Čech complex, because we don't know distances between sensors, just the information we can glean from the coverage/transmit data.

Nevertheless, this rather scant information is enough to answer the coverage question. First, we construct the *network graph* Γ, with sensors as vertices, and an edge between vertices if the sensors are within transmit distance. For three sensors which are in pairwise communication, the triangle spanned by the sensors is obviously contained in the cover. In Exercise 4.4.15, you will show that when the pairwise distances are exactly r_b, then the corresponding triangle of side length r_b is in the cover only if $r_c \geq \frac{r_b}{\sqrt{3}}$. Let R_Γ be the Rips complex of the network graph, and F_Γ the Rips complex of the fence.

Theorem 4.4.13 *For a set of sensors in D satisfying the assumptions above, there is complete coverage of D if there is*

$$\alpha \in H_2(R_\Gamma, F_\Gamma) \text{ with } \partial(\alpha) \neq 0.$$

Proof Map $R_\Gamma \xrightarrow{\sigma} D$ via

$$\tau \in (R_\Gamma)_k \mapsto \mathrm{Conv}(v_{i_0}, \ldots, v_{i_k}),$$

where τ corresponds to sensors located at positions $v_{i_0}, \ldots, v_{i_k}$.

By Lemma 4.3.4 the map σ induces a map σ_* on the relative homology defined in Sect. 4.2, yielding the commuting diagram below, where δ is the connecting homomorphism in the long exact sequence.

$$
\begin{array}{ccc}
H_2(R_\Gamma, F_\Gamma) & \xrightarrow{\ \delta\ } & H_1(F_\Gamma) \\
\downarrow{\scriptstyle \sigma_*} & & \downarrow{\scriptstyle \sigma_*} \\
H_2(D, \partial D) & \xrightarrow{\ \delta\ } & H_1(\partial D).
\end{array}
$$

Suppose there is not complete coverage, so there is $p \in D \setminus U$. This means the map σ factors thru $D \setminus p$, yielding the commuting diagram below:

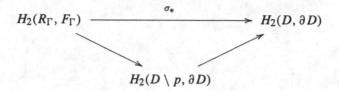

We can retract $D \setminus p$ to ∂D, so $H_2(D \setminus p, \partial D) = 0$, which forces $\sigma_* = 0$, so

$$0 = \delta \circ \sigma_* = \sigma_* \circ \delta. \tag{4.4.1}$$

But clearly $H_1(F_\Gamma) \xrightarrow{\sigma_*} H_1(\partial D)$ is an isomorphism. The connecting map

$$H_2(R_\Gamma, F_\Gamma) \xrightarrow{\delta} H_1(F_\Gamma)$$

is given by the snake lemma, and sends α to $\partial(\alpha)$. Therefore the assumption that $\partial(\alpha) \neq 0$ in $H_1(F_\Gamma)$ implies

$$\sigma_* \circ \delta(\alpha) \neq 0,$$

which contradicts Eq. 4.4.1. For further exposition, see [54].   □

Exercise 4.4.14 Prove that a configuration of sensors provides covering for all of D when the fence class $\gamma = \partial(D)$ is zero in homology.   ◇

Exercise 4.4.15 Prove that when the pairwise distances are exactly r_b, then the corresponding triangle of side length r_b is in the cover only if $r_c \geq \frac{r_b}{\sqrt{3}}$.   ◇

Chapter 5
Homology II: Cohomology to Ranking Problems

In this chapter we introduce cohomology; the constructions have roots in topology and geometry, and as such, can come with extra structure. We begin in Sect. 5.1 with Simplicial, Čech, and de Rham cohomology; a first difference from homology is that simplicial cohomology has a ring structure.

Building on de Rham cohomology, Sect. 5.2 begins with an overview of Hodge theory, which is a central (and technical) area of algebraic geometry. This is to set the stage to describe work of Jiang–Lim–Yao–Yuan in [94] which applies Hodge theory to ranking–specifically, the *Netflix Problem*, where the goal is to produce a coherent ranking from partial and inconsistent survey results, such as viewer ranking of movies. In particular, some rankings may involve loops: $a > b > c > a$. We revisit the Laplacian (Chap. 3) and describe the Hodge decomposition in a setting which only involves linear algebra.

In Sect. 5.3, we introduce CW complexes and cellular homology; the building blocks are *cells* and allow more flexibility than simplicial complexes in construction of spaces. In Sect. 5.4 we tackle duality, focusing on the theories of Poincaré and Alexander. Poincaré duality is like the Tai Chi symbol–the Yin and Yang are symmetric mirrors; see [45] for use in TDA. Alexander duality is like a black and white image: knowing one color tells us the other. TDA applications appear in [1] and [82].

- Cohomology: Simplicial, Čech, de Rham theories.
- Ranking and Hodge Theory.
- CW Complexes and Cellular Homology.
- Poincaré and Alexander Duality: Sensor Networks revisited.

© The Author(s), under exclusive license to Springer Nature Switzerland AG 2022
H. Schenck, *Algebraic Foundations for Applied Topology and Data Analysis*,
Mathematics of Data 1, https://doi.org/10.1007/978-3-031-06664-1_5

5.1 Cohomology: Simplicial, Čech, de Rham Theories

For a complex

$$\mathscr{C} : 0 \longrightarrow C^0 \xrightarrow{d^0} C^1 \xrightarrow{d^1} \cdots$$

we have $\mathrm{im}(d^i) \subseteq \ker(d^{i+1})$, and the *cohomology* of $\mathscr{C}$ is defined as

$$H^i(\mathscr{C}) = H^i\left(0 \longrightarrow C^0 \xrightarrow{d^0} C^1 \xrightarrow{d^1} \cdots\right)$$

So cohomology is homology with increasing indices. However, in a number of common situations, there is a multiplicative structure on the set of C^i. This results in a multiplicative structure on cohomology, yielding a ring $H^*(\mathscr{C}) = \bigoplus_i H^i(\mathscr{C})$.

5.1.1 Simplicial Cohomology

For a simplicial complex Δ and coefficient ring R, we have the complex of R-modules

$$\mathscr{C}_\bullet = \cdots \longrightarrow C_{i+1}(\Delta, R) \xrightarrow{d_{i+1}} C_i(\Delta, R) \xrightarrow{d_i} C_{i-1}(\Delta, R) \longrightarrow \cdots$$

Applying $\mathrm{Hom}_R(\bullet, R)$ and defining $C^i(\Delta, R) = \mathrm{Hom}_R(C_i(\Delta), R)$ yields

$$\mathscr{C}^\bullet = \cdots \longleftarrow C^{i+1}(\Delta, R) \xleftarrow{d^i} C^i(\Delta, R) \xleftarrow{d^{i-1}} C^{i-1}(\Delta, R) \longleftarrow \cdots$$

Elements of C_i are called i-chains, and elements of C^i are called i-cochains.

Definition 5.1.1 The multiplicative structure (called cup product) on simplicial cohomology is induced by a map on the cochains:

$$C^i(\Delta) \times C^j(\Delta) \xrightarrow{\cup} C^{i+j}(\Delta).$$

To define such a map we must define an action of a pair $(c_i, c_j) \in C^i(\Delta) \times C^j(\Delta)$ on an element of $C_{i+j}(\Delta)$: if $[v_0, \ldots v_{i+j}] = \sigma \in C_{i+j}(\Delta)$, then

$$(c_i, c_j)[v_0, \ldots v_{i+j}] = c_i[v_0, \ldots, v_i] \cdot c_j[v_i, \ldots, v_{i+j}],$$

where the last $\cdot$ is multiplication in the ring R.

Exercise 5.1.2 Compute the simplicial cohomology rings for the complexes $X \cap Y$ and $X \cup Y$ of Example 4.4.2, with coefficients in a field $\mathbb{K}$. Topologically,

$$X \cap Y \sim S^1 \text{ and } X \cup Y \sim S^2,$$

so you should find $H^1(X \cup Y) = 0$ and

$$H^0(X \cap Y) \simeq \mathbb{K} \simeq H^1(X \cap Y) \text{ and } H^0(X \cup Y) \simeq \mathbb{K} \simeq H^2(X \cup Y).$$

The interesting part of the exercise is determining the multiplicative structure. ◇

Exercise 5.1.3 For the ambitious: consult a book on algebraic topology and find a triangulation of the torus T^2 (the surface of a donut). Show the cohomology ring with $\mathbb{K}$ coefficients is $\mathbb{K}[x, y]/\langle x^2, y^2 \rangle$. Now do the same thing for the real projective plane $\mathbb{RP}^2$, first with $\mathbb{K} = \mathbb{Q}$ and then with $\mathbb{K} = \mathbb{Z}/2$ coefficients. You should get different answers; in particular *the choice of field matters*. This stems from the fact that the real projective plane is not orientable. ◇

5.1.2 Čech Cohomology

For a sheaf $\mathscr{F}$, Čech cohomology $\check{H}^i(\mathscr{F})$ is constructed so that the global sections of $\mathscr{F}$ defined in Sect. 4.3 of Chap. 3 are given by the zeroth cohomology. Let $\mathscr{U} = \{U_i\}$ be an open cover of X. If $\mathscr{U}$ consists of a finite number of open sets, then the lth module $\mathscr{C}^i$ in the Čech complex is simply

$$C^i(\mathscr{F}, \mathscr{U}) = \bigoplus_{\{j_0 < \ldots < j_i\}} \mathscr{F}(U_{j_0} \cap \cdots \cap U_{j_i}).$$

In general a cover need not be finite; in this case it is convenient to think of an element of $\mathscr{C}^i$ as an operator c_i which assigns to each $(i + 1)$-tuple $(j_0, \ldots, j_i)$ an element of $\mathscr{F}(U_{j_0} \cap \cdots \cap U_{j_i})$. We build a complex $\mathscr{C}^\bullet$ via:

$$\mathscr{C}^i = \prod_{\{j_0 < \ldots < j_i\}} \mathscr{F}(U_{j_0} \cap \cdots \cap U_{j_i}) \xrightarrow{d^i} \mathscr{C}^{i+1} = \prod_{\{j_0 < \ldots < j_{i+1}\}} \mathscr{F}(U_{j_0} \cap \cdots \cap U_{j_{i+1}}),$$

where $d^i(c_i)$ is defined by how it operates on $(i + 2)$-tuples, which is:

$$d^i(c_i)(j_0, \ldots, j_{i+1}) = \sum_{k=0}^{i+1} (-1)^k c_i(j_0, \ldots, \hat{j}_k, \ldots, j_{i+1})|_{\mathscr{F}(U_{j_0} \cap \cdots \cap U_{j_{i+1}})}.$$

Definition 5.1.4 The ith Čech cohomology of $\mathcal{F}$ and $\mathcal{U}$ is $\check{H}^i(\mathcal{U}, \mathcal{F}) = \check{H}^i(\mathcal{C}^\bullet)$.

Exercise 5.1.5 The constant sheaf on a space X is defined by giving $\mathbb{Z}$ the discrete topology, and defining $\mathbb{Z}(U)$ as the space of continuous functions from U to $\mathbb{Z}$.

(a) Is it true that $\mathbb{Z}(U) \simeq \mathbb{Z}$ for any open set U? If not, what additional assumption on U would make this true?

(b) Compute the Čech cohomology of the constant sheaf $\mathbb{Z}$ on S^2 using the cover of the open top hemisphere, and the two open bottom "quarterspheres" (all opens slightly enlarged so that they overlap) and see what you get. Your chain complex should start with these three opens (each of which is topologically an $\mathbb{R}^2$), then the three intersections (each of which is again an $\mathbb{R}^2$) and then the triple intersection, which is two disjoint $\mathbb{R}^2$'s.

(c) For the simplicial complex of Example 4.4.2 write down the Čech complex corresponding to open sets which are the triangular faces, slightly enlarged so they overlap. Compare to the simplicial homology.

(d) Find the Čech complex for the open cover of S^2 corresponding to the top hemisphere and bottom hemisphere, and compute the Čech cohomology. Why is your answer different from part (c)?

(e) Let X be a topological space with finite triangulation Δ, and create an open cover $\mathcal{U}$ by taking ϵ-enlarged neighborhoods of each face of Δ. Suppose that for every intersection $U = \cap_i U_i$ of open sets, U is contractible. Then $\check{H}^i(U, \mathbb{Z}) = 0$ for all $i \geq 1$, and Čech cohomology with $\mathbb{Z}$-coefficients agrees with the simplicial cohomology $H^i(\Delta, \mathbb{Z})$.

A cover as in part (e) is known as a *Leray cover*. See Theorem 9.4.5.   ◇

The definition of Čech cohomology is concrete and computable for a given cover, but part (d) of Exercise 5.1.5 illustrates a problem: different covers yield different results. This can be fixed by the following construction:

Definition 5.1.6 Order the set of open covers of X via $\mathcal{V} \preceq \mathcal{U}$ if $\mathcal{U}$ is finer than $\mathcal{V}$: for any $V \in \mathcal{V}$ there is a $U \in \mathcal{U}$ with $U \subseteq V$. This gives the set of all covers the structure of a poset, hence we can take the direct limit as in Definition 3.3.8. The ith Čech cohomology of $\mathcal{F}$ is

$$\check{H}^i(\mathcal{F}) = \varinjlim_{\mathcal{U}} H^i(\mathcal{U}, \mathcal{F})$$

While this is not a useful definition from a computational standpoint, the criterion of Leray appearing above gives sufficient conditions for a cover to compute $\check{H}^i(\mathcal{F})$. If all intersections of the open sets of a cover have no cohomology except at the zeroth position, then

$$\check{H}^i(\mathcal{F}) = \check{H}^i(\mathcal{U}, \mathcal{F})$$

For sheaves of modules $\mathscr{F}_i$ on a smooth variety X, we have the following key theorem:

Theorem 5.1.7 *Given a short exact sequence of sheaves*

$$0 \longrightarrow \mathscr{F}_1 \longrightarrow \mathscr{F}_2 \longrightarrow \mathscr{F}_3 \longrightarrow 0,$$

there is a long exact sequence in sheaf cohomology

$$\cdots \longrightarrow H^{i-1}(\mathscr{F}_3) \longrightarrow H^i(\mathscr{F}_1) \longrightarrow H^i(\mathscr{F}_2) \longrightarrow H^i(\mathscr{F}_3) \longrightarrow H^{i+1}(\mathscr{F}_1) \longrightarrow \cdots.$$

The proof rests on an alternate approach to defining sheaf cohomology, treated in Chap. 9. Here is a quick sketch: the global section functor is left exact and covariant. Using the derived functor approach of Chap. 9, we can take an injective resolution of $\mathscr{O}_X$-modules. By the derived functor machinery, a short exact sequence of sheaves yields a long exact sequence in the higher derived functors of H^0. Showing that these higher derived functors agree with the Čech cohomology would suffice to prove the theorem. This (along with the existence of an injective resolution) requires some work, and can be found in Section III.4 of [90]. The most important instance of a short exact sequence of sheaves is the ideal sheaf sequence. For a variety X sitting inside $\mathbb{P}^n$, say $X = V(I)$, we have the usual exact sequence of modules:

$$0 \longrightarrow I \longrightarrow R \longrightarrow R/I \longrightarrow 0.$$

Since exactness is measured on stalks, if we take an exact sequence of modules and look at the associated sheaves, we always get an exact sequence of sheaves. So there is an exact sequence of sheaves on $\mathbb{P}^n$:

$$0 \longrightarrow \mathscr{I}_X \longrightarrow \mathscr{O}_{\mathbb{P}^n} \longrightarrow \mathscr{O}_X \longrightarrow 0.$$

5.1.3 de Rham Cohomology

We saw in Chap. 3 that sheaves are built to encode local information, along with gluing data that allows us to translate on overlaps between local charts. A quintessential example is the sheaf of differential p-forms on a differentiable manifold X: for a sufficiently small open set U diffeomorphic to an open subset of $\mathbb{R}^n$ and local coordinates $\{x_1, \ldots, x_n\}$ on U,

$$\Omega_X^p(U) = \{f(x_1, \ldots, x_n) \cdot dx_{i_1} \wedge \cdots \wedge dx_{i_p}\},$$

where $f(x_1, \ldots, x_n)$ is a C^∞ function on U. For $p = 0$ we have

$$\Omega_X^0(U) = C^\infty(U).$$

These are familiar objects from vector calculus: on a surface S we integrate

$$f(s, t)ds \cdot dt,$$

which is an element of Ω_S^2.

The geometry of integration serves as a guide to what is going on here. Differentials measure volume: a two-form $ds \cdot dt$ represents an infinitesimal rectangle with sides ds and dt. The area of a parallelogram in $\mathbb{R}^2$ with sides given by vectors $\mathbf{v}$ and $\mathbf{w}$ is measured by the determinant, which is essentially what the two-form represents. Orientation matters:

$$\det \begin{bmatrix} v_1 & w_1 \\ v_2 & w_2 \end{bmatrix} = -\det \begin{bmatrix} w_1 & v_1 \\ w_2 & v_2 \end{bmatrix}.$$

When we take orientation into account (as opposed to just the area $ds \cdot dt$), we write $ds \wedge dt$, with the relation

$$ds \wedge dt = -dt \wedge ds,$$

coming from the determinant. This is the key to the main theorem of vector calculus: Stokes' theorem generalizes to higher dimensions, and the intuition remains the same: if $\omega \in \Omega_S^p$ and S is a closed, bounded $p + 1$ dimensional region, then the boundary $\partial(S)$ is p-dimensional, and under suitable conditions

$$\int_S d\omega = \int_{\partial(S)} \omega.$$

Exercise 5.1.8 For $X = \mathbb{R}^3$, differentiation takes a function $f(x_1, x_2, x_3)$ to the gradient ∇f, which is a one-form:

$$\Omega_X^0 \xrightarrow{d} \Omega_X^1 \text{ via } f \mapsto \sum_{i=1}^{3} \frac{\partial f}{\partial x_i} dx_i.$$

There are similar interpretations for $d\omega$ when $\omega \in \Omega_X^1$ or Ω_X^2 in terms of curl and divergence. Try working them out; if you get stuck, Spivak [138] will help.   ◇

Differentiation takes p-forms to $p + 1$-forms, hence there is a sequence

$$\cdots \longrightarrow \Omega_X^p \xrightarrow{d} \Omega_X^{p+1} \longrightarrow \cdots$$

Let

$$I_p = \{1 \le j_1 < j_2 < \cdots < j_p \le n\},$$

and for $P \in I_p$ let d_P denote the p-form $dx_{j_1} \wedge \cdots \wedge dx_{j_p}$. The key is that differentiating a p-form twice yields zero:

$$d^2\left(\sum_{P \in I_p} f_P d_P\right) = 0.$$

Exercise 5.1.9 Prove this as follows: fix a $\omega = f \cdot dx_{j_1} \wedge \cdots \wedge dx_{j_p}$, and examine the coefficients which appear in $d^2(\omega)$. This is the higher dimensional version of a result from vector calculus: for a vector field F

$$\mathrm{div}(\mathrm{curl}(F)) = 0.$$

The proof should remind you (exactly the same bookkeeping is involved) of the proof that the simplicial boundary operator ∂ satisfies $\partial^2 = 0$. To emphasize the fact that the order of indexing the p-forms matters, the operation of differentiation of p-forms is called *exterior differentiation*.   ◇

The *exterior algebra* provides a formal framework that is the right setting in which to investigate exterior differentiation and differential forms:

Definition 5.1.10 Let $V \simeq \mathbb{K}^n$ be a finite dimensional vector space. In Definition 2.2.15 of Chap. 2 the exterior algebra $\Lambda(V)$ was defined as a quotient of the tensor algebra $T(V)$ by the relations

$$\{v_i \otimes v_j + v_j \otimes v_i\}.$$

A quotient space has many guises; one way to obtain a concrete realization is to choose a basis $\{e_1, \ldots, e_n\}$ for V. Then $\Lambda(V) = \bigoplus_{i=0}^{n} \Lambda^i(V)$ is a 2^n-dimensional graded vector space. The relation $e_i \wedge e_j + e_j \wedge e_i = 0$ propagates to yield

$$e_{i_1} \wedge \cdots \wedge e_{i_p} = (-1)^{\mathrm{sgn}(\sigma)} e_{\sigma(i_1)} \wedge \cdots \wedge e_{\sigma(i_p)} \text{ for } \sigma \in S_n.$$

Thus, a basis for the pth graded component $\Lambda^p(V)$ of $\Lambda(V)$ is given by

$$\{e_{i_1} \wedge \cdots \wedge e_{i_p} \mid 1 \le i_1 < i_2 < \cdots < i_p \le n\}, \text{ with } \Lambda^0(V) = \mathbb{K}.$$

Example 5.1.11 For $V = \mathbb{K}^3$, a basis for $\Lambda(V)$ is

$$\{1, e_1, e_2, e_3, e_1 \wedge e_2, e_1 \wedge e_3, e_2 \wedge e_3, e_1 \wedge e_2 \wedge e_3\}.$$

Multiplication in $\Lambda(V)$ is called *wedge product* and written as $\wedge$. To determine the appropriate sign we must count the number of transpositions. So for example

$$e_2 \wedge (e_1 \wedge e_3) = -e_1 \wedge e_2 \wedge e_3 \text{ whereas } e_3 \wedge (e_1 \wedge e_2) = e_1 \wedge e_2 \wedge e_3.$$

The exterior algebra on $V \simeq \mathbb{K}^n$ behaves *almost* like the polynomial ring in n variables: in the polynomial ring, $e_i e_j = e_j e_i$, while in $\Lambda(V)$, $e_i e_j = -e_j e_i$. Hence $e_i^2 = 0$ for all $i \in \{1, \ldots, n\}$, and all monomials in $\Lambda(V)$ are squarefree.

Exercise 5.1.12 Write out the multiplication table for $\Lambda(\mathbb{K}^3)$, ignoring the identity. What can you say about the relation between the symmetric and exterior algebras when the field $\mathbb{K}$ has characteristic two? $\diamond$

There is a point to make here: the exterior algebra of a vector space $V \simeq \mathbb{K}^n$ is constructed in a purely formal fashion. If $R = \mathbb{K}[x_1, \ldots, x_m]$ and we replace V with the free module R^n, the same increase in complexity that occurred in passing from vector spaces to modules manifests itself. In particular, we can differentiate polynomials, so $\Lambda(R^n)$ comes with the additional operation of exterior differentiation, whereas $\Lambda(\mathbb{K}^n)$ has no such operation. It is by passing to a more complicated algebraic object that we make it possible for geometry to enter the picture.

Definition 5.1.13 Exercise 5.1.9 shows that applying the exterior differentiation operator twice yields zero. Georges de Rham used this insight to define an eponymous cohomology theory:

$$H^p_{dR}(X) = H^p \left(\cdots \longrightarrow \Omega^p_X \xrightarrow{\ d\ } \Omega^{p+1}_X \longrightarrow \cdots \right)$$

We seem to have strayed very far afield from the comfortable combinatorial setting of simplicial complexes and simplicial homology. But this is not the case at all!

Theorem 5.1.14 (de Rham Theorem) *Let X be a smooth manifold, and Δ a triangulation of X. Then the simplicial cohomology (with real coefficients) and de Rham cohomology groups are isomorphic:*

$$H^p_{dR}(X) \simeq H^p(\Delta, \mathbb{R}).$$

Proof (Sketch) As in the proof of the snake lemma, the key is to define the right map, which we do at the level of cochains. A map

$$C^p_{dR} \longrightarrow C^p(\Delta),$$

takes as input a p-form ω, and produces an element of $\mathrm{Hom}_\mathbb{R}(C_p(\Delta), \mathbb{R})$. Let $\sigma \in C_p(\Delta)$. Since integration is a linear operator, we have a pairing

$$\langle \omega, \sigma \rangle = \int_\sigma \omega.$$

It is not an isomorphism at the level of chains, but it is at the level of cohomology. For a proof of this, and of the fact that any smooth manifold can be triangulated, see [135]. $\square$

5.2 Ranking, the Netflix Problem, and Hodge Theory

Hodge theory is a central area of geometry, focused on the study of differential forms; see Voisin [147] for a comprehensive treatment. This section describes an application of Hodge theory to data science. The Netflix problem asks for a global ranking of objects, when individual input ranking data is inconsistent. In particular, the rankings provided as input may contain cycles $a > b > c > a$. We start off by giving an overview of Hodge theory, then derive a structure theorem (the Hodge decomposition) in the simplest possible setting. We close with a sketch of the approach of [94] to the Netflix problem via the Hodge decomposition.

5.2.1 Hodge Decomposition

In the setting of the previous section, we have

Theorem 5.2.1 (Hodge Theorem) *For a smooth compact Riemannian manifold X, there is a decomposition*

$$\Omega_X^k \simeq \operatorname{im}(d^{k-1}) \oplus \operatorname{im}(d^{k^*}) \oplus \ker(L)$$

where L is the Laplacian appearing in Eq. 5.2.1 below, and

$$H_{dR}^k(X) \simeq \ker(L).$$

Riemannian means there is a positive definite inner product on the tangent space of X. Before we unpack the terms in the formula, we highlight the main consequence: homology and cohomology are represented as quotient spaces, so there is *almost never* any hope of finding canonical representatives. The Hodge theorem says this is not the case for X satisfying the hypotheses of Theorem 5.2.1: the kernel of the Laplacian provides exactly such canonical representatives. The operator d^{k-1} in the formula above is exterior differentiation, the operator d^{k^*} is the *adjoint* operator

$$\Omega_X^{k+1} \xrightarrow{d^{k^*}} \Omega_X^k,$$

and the Laplacian L is defined as the operator

$$d^{k^*} \circ d^k + d^{k-1} \circ d^{k-1^*} \tag{5.2.1}$$

The first term $d^{k^*} \circ d^k$ of the Laplacian "goes forward then comes back", while the second term $d^{k-1} \circ d^{k-1^*}$ "goes backwards then goes forwards". An important ingredient in the proof (which uses elliptic PDE, and is beyond the scope of these notes) is the Hodge star operator. In the case of the exterior algebra on $V \simeq \mathbb{K}^n$, we have

$$\Lambda V \simeq \bigoplus \Lambda^i(V)$$

and Hodge star identifies $\Lambda^k(V)^{\perp}$ with $\Lambda^{n-k}(V)$ via the pairing

$$\Lambda^k(V) \times \Lambda^{n-k}(V) \longrightarrow \Lambda^n(V) \simeq \mathbb{K} \text{ via } (v, w) \mapsto v \wedge w.$$

Stripping down to the bare essentials of the inner product space setting, we next give a lowbrow proof of the Hodge decomposition.

Definition 5.2.2 Let V, W be finite dimensional inner product spaces, and

$$V \xrightarrow{A} W$$

a linear transformation. For concreteness, let $\mathbb{K} = \mathbb{R}$, with inner product given by the usual dot product, and choose bases so that A is a matrix. The adjoint operator A^* is defined via

$$\langle A\mathbf{v}, \mathbf{w} \rangle = \langle \mathbf{v}, A^*\mathbf{w} \rangle.$$

So in this setting, we have

$$\langle A\mathbf{v}, \mathbf{w} \rangle = \mathbf{w}^T \cdot A\mathbf{v} = (A\mathbf{v})^T \cdot \mathbf{w} = \mathbf{v}^T \cdot A^T \mathbf{w} = \langle \mathbf{v}, A^*\mathbf{w} \rangle.$$

Proposition 5.2.3 *Let*

$$V_1 \xrightarrow{d_1} V_2 \xrightarrow{d_2} V_3$$

be a complex of finite dimensional inner product spaces. Then

$$V_2 \simeq \mathrm{im}(d_1) \oplus \mathrm{im}(d_2^T) \oplus \ker(L), \text{ where } L = d_1 d_1^T + d_2^T d_2.$$

Proof Let $\mathrm{rank}(V_i) = a_i$, $\mathrm{rank}(d_i) = r_i$, $\dim \ker(d_i) = k_i$. Choose bases so that

$$d_1 = \begin{bmatrix} I_{r_1} & 0 \\ 0 & 0 \end{bmatrix} \text{ and } d_2 = \begin{bmatrix} 0 & 0 \\ 0 & I_{r_2} \end{bmatrix}$$

where 0 represents a zero matrix of the appropriate size. For example, since the matrix d_1 is $a_2 \times a_1 = a_2 \times (r_1 + k_1)$, the top right zero in d_1 has r_1 rows and k_1 columns. Then we have that $d_1 d_1^T$ and $d_2^T d_2$ are both $a_2 \times a_2$ matrices, with

$$d_1 d_1^T = \begin{bmatrix} I_{r_1} & 0 \\ 0 & 0 \end{bmatrix} \text{ and } d_2^T d_2 = \begin{bmatrix} 0 & 0 \\ 0 & I_{r_2} \end{bmatrix}.$$

Hence

$$L = d_1 d_1^T + d_2^T d_2 = \begin{bmatrix} I_{r_1} & 0 & 0 \\ 0 & 0 & 0 \\ 0 & 0 & I_{r_2} \end{bmatrix}.$$

The zero in position $(2, 2)$ represents a square matrix of size

$$k_2 - r_1 = \dim(\ker(d_2)/\mathrm{im}(d_1)),$$

and the linear algebra version of the Hodge decomposition follows. □

5.2.2 Application to Ranking

In [94], Jiang–Lim–Yao–Ye apply the Hodge decomposition to the Netflix problem. Viewers rank movies; producing a global ranking is complicated by the presence of loops. We begin by creating a weighted directed graph G, where a weight of m on the directed edge $[a, b]$ represents an aggregate of m voters preferring movie $[a]$ to movie $[b]$. Construct a two dimensional simplicial complex Δ_G by adding in the face $[abc]$ if all edges of the triangle are present. This is a flag complex, as in Definition 4.4.8 in Chap. 4.

The assignment of weights to each edge defines a linear functional on $C_1(\Delta_G)$, which by Proposition 5.2.3 has a decomposition

$$C_1(\Delta_G, \mathbb{R}) \simeq \mathrm{im}(d_1^T) \oplus \mathrm{im}(d_2) \oplus \ker(L).$$

Cycles of any length represent inconsistencies, so the global ranking should ignore them. Cycles of length three have been filled in to create two-simplices, which are represented by $\mathrm{im}(d_2)$. Cycles of length four or more correspond to homology classes, so are represented by $\ker(L)$. Hence we have

Theorem 5.2.4 (Jiang–Lim–Yao–Ye) *For inconsistent ranking data as in the Netflix problem, orthogonal projection onto the subspace*

$$\mathrm{im}(d_1^T)$$

produces the global ranking which most closely reflects the voter preferences.

Proof The Hodge decomposition. □

Exercise 5.2.5 Relate the Hodge decomposition to SVD (Chap. 1, Sect. 4.3). ◇

Since we mentioned flag complexes above, it seems worth stating a famous open question on flag complexes: the *Charney–Davis* [38] conjecture.

Conjecture 5.2.6 Let Δ be a flag complex which triangulates a sphere S^{2d-1}, and let f_{i-1} denote the number of $i - 1$ dimensional faces of Δ. Then

$$(-1)^d \sum_{i=0}^{2d} \left(\frac{-1}{2}\right)^i f_{i-1} \geq 0.$$

Davis–Okun prove the $d = 2$ case in [49] using ℓ^2-cohomology.

5.3 CW Complexes and Cellular Homology

CW complexes are less intuitive than simplicial or singular complexes. The utility of CW complexes is that they provide a more economical way to construct spaces than can be done when using other methods, and this manifests in simpler homology computations. A topological space is *Hausdorff* if for any pair of points $p_1 \neq p_2$ there exist disjoint open sets U_1 and U_2 with $p_i \in U_i$ such that

$$U_1 \cap U_2 = \emptyset.$$

A CW complex is a Hausdorff topological space built inductively by attaching open balls B^n to a lower dimensional scaffolding X^{n-1} in a prescribed way. A zero dimensional CW complex X^0 is a discrete set of points, and a one-dimensional CW complex is built by gluing copies of the unit interval $[0, 1]$ to X^0, with the requirement that $\partial[0, 1] = \{0, 1\} \mapsto X^0$. In general, we have

Definition 5.3.1 An n-dimensional CW complex is the union of

- Open balls B^n, known as *cells*.
- An $n - 1$ dimensional CW complex X^{n-1}, the $n - 1$ *skeleton*,

quotiented by the image of continuous attaching maps for each ball B^n:

$$\partial(\overline{B^n}) = S^{n-1} \mapsto X^{n-1}.$$

Example 5.3.2 The n-sphere S^n may be constructed with only two cells: an n-cell B^n and zero cell p, with gluing map collapsing $\partial(\overline{B^n}) = S^{n-1}$ to the point p. The *algebraic* chain complex we construct has nonzero terms only in positions n and 0, and all differentials are zero, allowing computation of the homology of S^n with no effort at all. Other examples of CW complexes:

- Simplicial and Polyhedral Complexes.
- Projective Space and Grassmannians.
- Compact Surfaces.
- Smooth Manifolds*

Of course, the classes above overlap; the asterisk for manifolds reflects the fact that there is a CW complex which is homotopic to a given manifold.

CW complexes were introduced by J.H.C. Whitehead; the nomenclature arises from the fact that a CW complex X

- is closure finite (C): the closure of a cell of X intersects only a finite number of other cells of X.
- has the weak topology (W): $Y \subseteq X$ is closed iff $Y \cap \bar{e}$ is closed for every cell e of X.

To define homology for CW complexes, we need *relative homology*, which appeared in Sect. 4.2 of Chap. 4. Suppose $X \subseteq Y$ are topological spaces; for simplicity let X be a simplicial complex which is a subcomplex of Y. Then for each k we have short exact sequences

$$0 \longrightarrow C_k(X) \overset{i}{\longrightarrow} C_k(Y) \overset{j}{\longrightarrow} C_k(Y)/C_k(X) \longrightarrow 0.$$

The maps i and j commute with the boundary operators, and Theorem 4.3.6 of Chap. 4 shows this yields a long exact sequence

$$\cdots \to H_k(X) \overset{i_*}{\to} H_k(Y) \overset{j_*}{\to} H_k(Y, X) \overset{\delta}{\to} H_{k-1}(X) \to \cdots \qquad (5.3.1)$$

Here $H_k(Y, X)$ denotes the homology of the quotient complex whose k^{th} term is $C_k(Y)/C_k(X)$. We apply this to construct a chain complex from a CW complex X, using the k-skeleta X^k of X as building blocks.

Lemma 5.3.3 *For a CW complex X, define $D_p(X) = H_p(X^p, X^{p-1})$, and define*

$$D_p(X) \overset{\partial}{\longrightarrow} D_{p-1}(X)$$

via the composite map $\partial = j_* \delta$

$$H_p(X^p, X^{p-1}) \xrightarrow{\qquad \partial \qquad} H_{p-1}(X^{p-1}, X^{p-2})$$

with δ and j_* mapping through $H_{p-1}(X^{p-1})$

Then $\partial^2 = 0$.

Proof Because $\partial^2 = (j_* \delta) \cdot (j_* \delta)$, the result follows from

$$\delta j_* = 0,$$

which is a consequence of the exactness of the sequence in Eq. 5.3.1. Notice that since X^k is dimension k, $H_n(X^k) = 0$ for $n > k$.   □

Example 5.3.4 The torus T^2 and Klein bottle KB can both be constructed as a CW complex with a single two cell, a pair of one cells, and a single zero cell, as below

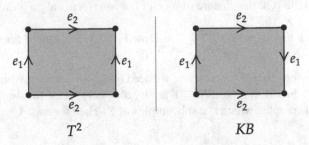

$$T^2 \qquad\qquad\qquad KB$$

Hence the cellular homology with $\mathbb{Z}$-coefficients is given for both spaces by

$$0 \longrightarrow \mathbb{Z}^1 \xrightarrow{\partial_2} \mathbb{Z}^2 \xrightarrow{\partial_1} \mathbb{Z}^1 \longrightarrow 0.$$

All differentials are zero, except ∂_2 of the Klein bottle, which sends the two cell to twice the vertical edge. So

	T^2	KB
H_0	$\mathbb{Z}$	$\mathbb{Z}$
H_1	$\mathbb{Z}^2$	$\mathbb{Z} \oplus \mathbb{Z}/2\mathbb{Z}$
H_2	$\mathbb{Z}$	0

5.4 Poincaré and Alexander Duality: Sensor Networks Revisited

Henri Poincaré discovered his duality theorem in 1893; it has an intrinsic appeal because it concerns symmetry. In Poincaré duality, there is a single underlying object (the manifold M), and the duality involves the homology of M. James Alexander proved his duality theorem in 1915; in contrast to Poincaré duality, Alexander duality concerns two different spaces. Throughout this section the coefficient ring R will be a field $\mathbb{K}$, and we sometimes write $H_i(*)$ for $H_i(*, \mathbb{K})$.

5.4.1 Statement of Theorems and Examples

Theorem 5.4.1 (Poincaré Duality) *Let M be a compact, connected, orientable manifold of dimension n without boundary. Then there is a perfect pairing*

$$H_k(M) \otimes H_{n-k}(M) \longrightarrow \mathbb{K}.$$

This says that there is a vector space isomorphism $\mathrm{Hom}_{\mathbb{K}}(H_{n-k}(M), \mathbb{K}) \simeq H_k(M, \mathbb{K})$, which is often phrased in terms of cohomology

$$H^k(M, \mathbb{K}) \simeq H_{n-k}(M, \mathbb{K}).$$

Recall from Sect. 5.1 that for a simplicial complex Δ,

$$C^i(\Delta, R) = \mathrm{Hom}_R(C_i(\Delta), R)$$

and that the simplicial cohomology $H^i(\mathscr{C})$ is the cohomology at the ith position. The dimension b_i of $H_i(\Delta, \mathbb{K})$ is called the ith *Betti number* of Δ. So Poincaré duality asserts that for M as in the statement of Theorem 5.4.1, the vector of Betti numbers $\mathbf{b}(M) = (b_0, \ldots, b_n)$ is a palindrome: $(b_0, \ldots, b_n) = (b_n, \ldots, b_0)$.

Example 5.4.2 Our computations earlier in the chapter show that

$$
\begin{aligned}
\mathbf{b}(S^1) &= &(1, 1) &\quad \text{by Example 4.2.5.}\\
\mathbf{b}(S^2) &= &(1, 0, 1) &\quad \text{by Example 4.2.8.}\\
\mathbf{b}(S^n) &= &(1, 0^{n-1}, 1) &\quad \text{by Example 5.3.2.}
\end{aligned}
$$

Example 5.3.4 showed that the Torus has Betti vector $(1, 2, 1)$, whereas the Klein Bottle has Betti vector $(1, 1, 0)$. We don't expect Poincaré duality to hold for the Klein Bottle, since it is not orientable. However, see Example 5.4.19.

Theorem 5.4.3 (Alexander Duality) *For a compact, locally contractible $X \subseteq S^n$, the homology of X and of $S^n \setminus X$ are related via*

$$\widetilde{H}_k(S^n \setminus X) \simeq \widetilde{H}^{n-1-k}(X)$$

where $\widetilde{H}^i$ denotes reduced *cohomology.*

In terms of the *reduced* Betti vector, this says that $\widetilde{\mathbf{b}}(S^n \setminus X)$ is obtained by reversing $\widetilde{\mathbf{b}}(X)$, then shifting left by one.

Example 5.4.4 Let $n = 2$, with S^2 triangulated as in Δ_1 of Example 4.2.8, and let X be the subcomplex with maximal faces $\{[1, 2], [2, 3], [3, 1]\}$, so X is homotopic to S^1. The homology of X was computed in Example 4.2.5. Appending zeros so the reduced Betti vector of X has length $n + 1 = 3$ yields

$$\widetilde{\mathbf{b}}(X) = (0, 1, 0).$$

What about $S^2 \setminus X$? Since singular and simplicial homology agree, we may use singular homology to compute. As singular homology is homotopy invariant, removing the $X = S^1$ "belt" from S^2 results in two disconnected disks, which are each homotopic to a point. Therefore $\mathbf{b}(S^2 \setminus X) = (2, 0, 0)$, hence

$$
\begin{aligned}
\widetilde{\mathbf{b}}(S^2 \setminus X) \qquad &= \quad (1, 0, 0) \\
\text{reverse } (1, 0, 0) \text{ to obtain} \qquad &\quad (0, 0, 1) \\
\text{shift } (0, 0, 1) \text{ left one step} \qquad &\quad (0, 1, 0) \quad = \quad \widetilde{\mathbf{b}}(X).
\end{aligned}
$$

5.4.2 Alexander Duality: Proof

We prove a combinatorial version of Alexander Duality, following the elegant argument of Björner–Tancer [17]. The setup is the following: Δ will be a simplicial complex on the vertex set $V = \{v_1, \ldots, v_n\}$. We write S for the simplicial complex on V whose maximal faces correspond to the subsets of V having $n - 1$ elements; therefore S is homotopic to S^{n-2}.

Definition 5.4.5 The combinatorial Alexander dual of Δ is

$$\Delta^* = \{\sigma \subset V \mid \overline{\sigma} \notin \Delta\}, \text{ where } \overline{\{v_{i_1}, \ldots, v_{i_j}\}} = \{v_1, \ldots, v_n\} \setminus \{v_{i_1}, \ldots, v_{i_j}\}.$$

The condition that $\overline{\sigma} \notin \Delta$ means that $\overline{\sigma}$ is not a face of Δ.

Example 5.4.6 For X in Example 5.4.4, the nonfaces of Δ are

$$\{[1, 2, 3], [1, 2, 4], [1, 3, 4], [2, 3, 4], [1, 4], [2, 4], [3, 4], [4]\},$$

so the complements of the nonfaces are

$$\{[4], [3], [2], [1], [2, 3], [1, 3], [1, 2], [1, 2, 3]\}.$$

Hence Δ^* consists of the vertex $[4]$ and the triangle $[1, 2, 3]$, so Δ^* is homotopic to a pair of points.

Since $|V| = n$, the corresponding sphere is an S^{n-2}, so combinatorial Alexander duality will take the form

$$\widetilde{H}_i(\Delta) \simeq \widetilde{H}^{n-3-i}(\Delta^*). \tag{5.4.1}$$

Lemma 5.4.7 *For $j \in \sigma \in \Delta$, $\mathrm{sgn}(j, \sigma) = (-1)^{i-1}$, where j is the ith smallest element in the set σ. Let*

$$p(\sigma) = \prod_{i \in \sigma} (-1)^{i-1}.$$

Then for $k \in \sigma$, $\mathrm{sgn}(k, \sigma)p(\sigma \setminus k) = \mathrm{sgn}(k, \overline{\sigma} \cup k)p(\sigma)$.

Exercise 5.4.8 Show both sides of the expression above equal $(-1)^{k-1}$. ◇

Let $e_\sigma \in \tilde{C}_i$ denote the element corresponding to $\sigma \in \Delta_i$, so that

$$\partial_i(e_\sigma) = \sum_{j \in \sigma} \mathrm{sgn}(j, \sigma)e_{\sigma \setminus j}. \tag{5.4.2}$$

Dualizing to the C^i, this means the dual maps in cohomology are given by

$$\partial^i(e_\sigma^*) = \sum_{\substack{j \notin \sigma \\ \sigma \cup j \in \Delta}} \mathrm{sgn}(j, \sigma \cup j)e_{\sigma \cup j}^*. \tag{5.4.3}$$

Equation 5.4.1 will follow from the following two isomorphisms, where 2^V is the full (solid) $n - 1$ simplex $[v_1, \ldots, v_n]$.

$$\tilde{H}_i(\Delta) \sim \tilde{H}_{i+1}(2^V, \Delta) \tag{5.4.4}$$

and

$$\tilde{H}_{i+1}(2^V, \Delta) \simeq \tilde{H}^{n-3-i}(\Delta^*). \tag{5.4.5}$$

Equation 5.4.4 follows from the long exact sequence in relative homology for the pair $\Delta \subseteq 2^V$, combined with all $\tilde{H}_i(2^V) = 0$. The vanishing of $\tilde{H}_i(2^V)$ for all i follows since 2^V is homotopic to a point. The key is showing Eq. 5.4.5 holds.

Proof (Combinatorial Alexander Duality–Equation 5.4.1) Write the differential in relative homology for the pair $(2^V, \Delta)$ as d_j and the relative chains as R_j; the operator d_j is given by essentially the same formula as Eq. 5.4.2, except that chains with image in $C_i(\Delta)$ are zero; accounting for this we have:

$$d_j(e_\sigma) = \sum_{\substack{k \in \sigma \\ \sigma \setminus k \notin \Delta}} \mathrm{sgn}(k, \sigma)e_{\sigma \setminus k}. \tag{5.4.6}$$

Write C^j for relative cochains and d^j for the relative coboundary operator, so that

$$d^j(e_\sigma^*) = \sum_{\substack{k \notin \sigma \\ \sigma \cup k \in \Delta^*}} \mathrm{sgn}(k, \sigma \cup k) e_{\sigma \cup k}^* = \sum_{\substack{k \in \overline{\sigma} \\ \overline{\sigma} \setminus k \notin \Delta}} \mathrm{sgn}(k, \sigma \cup k) e_{\overline{\sigma \setminus k}}^*. \qquad (5.4.7)$$

To complete the proof we bring the operator p from Lemma 5.4.7 into play. Define

$$R_j \xrightarrow{\phi_j} C^{n-j-2} \text{ via } \phi_j(e_\sigma) = p(\sigma) e_{\overline{\sigma}}^*.$$

Then the diagram

$$
\begin{array}{ccccccc}
\cdots & \longrightarrow & R^j & \xrightarrow{\ \ d_j\ \ } & R_{j-1} & \longrightarrow & \cdots \\
& & \downarrow{\scriptstyle \phi_j} & & \downarrow{\scriptstyle \phi_{j-1}} & & \\
\cdots & \longrightarrow & C^{n-2-j} & \xrightarrow{\ d^{n-1-j}\ } & C_{n-j-1} & \longrightarrow & \cdots
\end{array}
$$

commutes, and therefore induces the isomorphism of Eq. 5.4.1. □

Exercise 5.4.9 Check that indeed $\phi_{j-1} d_j = d^{n-1-j} \phi_j$. ◇

5.4.3 Sensor Networks Revisited

Example 4.4.12 described the work of de Silva-Ghrist on coverage of a region D by a sensor network *at a particular instant in time*. We now modify the problem: instead of a single snapshot in time, we have a number of snapshots taken over some time interval. For concreteness, suppose the time interval is $t \in [0, \dots, 1]$ and a finite set of snapshots of coverage are taken at times

$$0 = t_1 < t_2 < \cdots t_{m-1} < t_m = 1.$$

Can an intruder attempting to steal the diamonds in Example 4.4.12 evade detection? Following [52], we frame the problem as follows. Let $D \subseteq \mathbb{R}^d$ be a bounded domain homeomorphic to a ball, and suppose we have a set of sensors $S = \{v_1, \dots, v_n\}$, each providing coverage of a unit d-ball, wandering through D. As in Example 4.4.12, the sensors do not know coordinate information, but do know when their coverage region overlaps with another sensor. Define the network graph Γ_t at time t as in Example 4.4.12: vertices are sensors, with an edge connecting sensors if their coverage overlaps. Let

$$X(t) = \bigcup_{v_i(t) \in S} B_{v_i(t)} \text{ and } X = \bigcup_{t \in I} X(t) \times \{t\} \subseteq (D \times [0, 1]).$$

So $X(t)$ is the coverage of the sensor network at time t, and X is the subset of spacetime covered by sensors.

Definition 5.4.10 An evasion path is a section (Chap. 3) of the projection map

$$X^c = (D \times [0, 1]) \setminus X \xrightarrow{\pi} [0, 1].$$

So an evasion path is a map $[0, 1] \xrightarrow{s} X^c$ such that $s(t) \notin X(t)$ for any $t \in [0, 1]$.

Example 5.4.11 Let $X(t)$ and Γ_t be as below: snapshots are taken at $t \in \{0, \frac{1}{2}, 1\}$

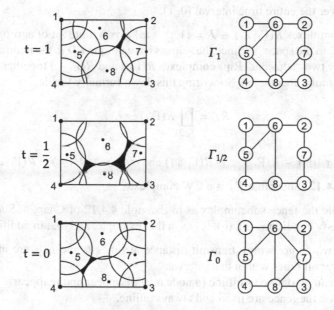

The space X itself is depicted below. Notice that there is no way for an intruder to escape detection–if they are not detected at $t = 0$, they are in the region $X(0)^c$, and that region shrinks continuously over time, finally closing out a bit after $t = \frac{1}{2}$.

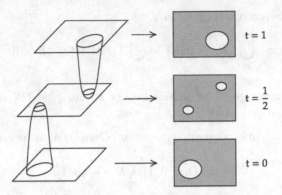

In particular, even though at every point in time there is a region where an intruder can escape detection (that is, each snapshot on the right hand side of the figure above has a gap in the coverage region), there is not a way for an intruder to escape detection over the entire time interval $[0, 1]$.

The Rips complexes $R(\Gamma_{t_j})$, $j \in J = \{1, \ldots, m\}$ give a finite set of approximations $X_1, \ldots, X_m$ to the space X. Since the sensors have unique ID's, the natural strategy is to identify two "adjacent" Rips complexes $R(\Gamma_{t_i})$ and $R(\Gamma_{t_{i+1}})$ together along the largest common subcomplex. So writing this down formally yields

$$R_J = \coprod_{j \in J} R(\Gamma_{t_j})/\sim,$$

where $\sigma \sim \tau$ iff $\sigma \in R(\Gamma_{t_j})$, $\tau \in R(\Gamma_{t_j+1})$ and $\sigma = \tau \in R(\Gamma_{t_j}) \cap R(\Gamma_{t_j+1})$.

Exercise 5.4.12 Show that R_J is a CW complex.   ◇

Let F denote the fence subcomplex as in Example 4.4.12 of Chap. 4. Suppose the sensors satisfy conditions (a), (b), (c) as in that example, and that in addition

- Sensors which are within transmit distance at times t_i and t_{i+1} are in transmit distance at all times within that interval.
- Sensors can go online or offline (a node may disappear and reappear).
- Sensors on the fence are fixed and always online.

Exercise 5.4.13 With the conditions above, prove that if there exists

$$\alpha \in H_2(R_J, F) \text{ such that } \partial(\alpha) \neq 0,$$

then there is no evasion path. Equivalently, the projection map π of Definition 5.4.10 has no section: there is not a continuous curve $C(t) : [0, 1] \longrightarrow X^c$.   ◇

In [1], Adams–Carlsson use Alexander duality to prove a necessary criterion for existence of an evasion path, and in [82], Ghrist–Krishnan give necessary and sufficient conditions for evasion in arbitrary dimension.

5.4.4   Poincaré Duality

Poincaré duality is in some sense the paradigm for duality theories in topology. It can be proved in a purely algebraic manner, via cap product with the fundamental class. A more easily visualized low-tech proof involves several fundamental tools such as barycentric subdivision and dual cell decomposition, and this is the proof we sketch below.

Definition 5.4.14 Let σ_n be a geometric n-simplex, so that σ_n is the convex hull of vertices $\{v_0, \ldots, v_n\}$. The *barycenter* $b(\sigma_n)$ of σ_n is

$$b(\sigma_n) = \frac{1}{n+1} \sum_{i=0}^{n} v_i.$$

The first *barycentric subdivision* $B(\sigma_n)$ of σ_n is obtained iteratively:

- B_1 is the the one-dimensional simplicial complex obtained by barycentrically subdividing the edges of σ_n.
- B_{i+1} is the subdivision of the $i+1$ skeleton of σ_n obtained by coning the barycenters of the $i+1$ simplices with B_i.

Example 5.4.15 The barycentric subdivision of a 2-simplex:

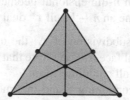

Exercise 5.4.16 (Poincaré–Hopf) A beautiful application of barycentric subdivision involves the Poincaré–Hopf theorem; we follow the treatment in §8 of [76]. Let $\mathcal{V}$ be a vector field on a smooth, compact orientable surface S of genus g–"a sphere with g handles". A point $p \in S$ is a *singularity* of $\mathcal{V}$ if $\mathcal{V}(p) = 0$. Fix a small disk D around p such that $\mathcal{V}$ has no other singularities on D (so $\mathcal{V}$ has only isolated singularities), and let $\partial(D) = \Gamma$, oriented counterclockwise. If $\mathcal{V}$ is locally defined by (f, g), the index of $\mathcal{V}$ at p is the integer $\mathrm{ind}_p(\mathcal{V}) = \frac{1}{2\pi} \int_\Gamma \frac{f\,dg - g\,df}{f^2 + g^2}$. The Poincaré–Hopf theorem is

$$\sum_{p \in \mathrm{sing}(\mathcal{V})} \mathrm{ind}_p(\mathcal{V}) = \chi(S). \tag{5.4.8}$$

This is amazing: a property of vector fields is governed by topology. Prove Eq. 5.4.8 for the following special class of vector fields.

Let Δ be a triangulation of S, and construct a vector field $\mathcal{V}$ on S as "flow towards the barycenters of the faces of Δ", depicted below for a two simplex:

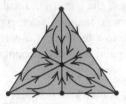

Since the barycenter of a vertex v is v itself, the singularities of $\mathcal{V}$ occur only at the barycenters of the faces of Δ. The index computation is local; translate so p is the origin in $\mathbb{R}^2$ and use vector calculus to compute that

$$\mathrm{ind}_p(\mathcal{V}) \;=\; 1 \qquad \text{if } p = b(v) \text{ for } v \text{ a vertex of } \Delta. \qquad \text{Hint: } \mathcal{V} = \langle x, y \rangle.$$
$$\mathrm{ind}_p(\mathcal{V}) \;=\; -1 \qquad \text{if } p = b(\epsilon) \text{ for } \epsilon \text{ an edge of } \Delta. \qquad \text{Hint: } \mathcal{V} = \langle y, x \rangle.$$
$$\mathrm{ind}_p(\mathcal{V}) \;=\; 1 \qquad \text{if } p = b(\tau) \text{ for } \tau \text{ a triangle of } \Delta. \qquad \text{Hint: } \mathcal{V} = \langle -x, -y \rangle.$$

Why does the result follow from this? Example 4.2.4 will be useful.   ◇

With barycentric subdivision in hand, we next describe the dual cell subdivision D of a simplicial complex Δ. The resulting object D is a CW complex.

Definition 5.4.17 Let Δ be an n-dimensional geometric simplicial complex, and let τ be a k-face of Δ. We define an $n - k$ cell τ^* dual to τ as follows.

- Form the first barycentric subdivision of Δ: the union of the first barycentric subdivision of each maximal face σ of Δ. Note that if Δ triangulates a manifold, the maximal faces of Δ are all equidimensional.
- For each maximal face σ containing τ and intermediate face τ' such that $\tau \subseteq \tau' \subseteq \sigma$, let $b_{\tau'}$ be the barycenter of τ'. The dual cell is defined by letting $\tau_\sigma = \tau^* \cap \sigma = \mathrm{Conv}\{b_{\tau'} \mid \tau \subseteq \tau' \subseteq \sigma\}$, and

$$\tau^* = \bigcup_{\tau \subseteq \sigma \in \Delta_n} \tau_\sigma.$$

Notice that $\tau^* \cap \tau$ meet in a unique point b_τ.

Consider two homology classes C and C' of complementary dimensions k and $n - k$ on an n-dimensional manifold. Cycles are equivalence classes, so they can be moved. One way to formulate Poincaré duality is that if cycles are moved so that C and C' meet transversely at a point p, then

$$T_p(X) = T_p(C) \oplus T_p(C').$$

See [84] for a proof along these lines. This is also essentially what happens with the dual cell decomposition.

Example 5.4.18 For the sphere S^2 with triangulation as in Example 4.2.8, the dual cell decomposition is

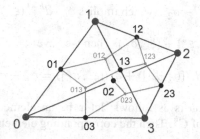

For an edge τ with vertices $\{i, j\}$, the barycenter b_τ is labelled ij, and similarly for triangles.

Proof (Sketch) The dual cell decomposition gives a $1 - 1$ map

$$C^k(\Delta) \longrightarrow C_{n-k}(D) = D_{n-k}.$$

The relation between the coboundary map $C^{k-1}(\Delta) \to C^k(\Delta)$ and the boundary map of the dual cell complex $D_{n-k} = C_{n-k}(D) \to C_{n-k-1}(D) = D_{n-k-1}$ is illustrated in Example 5.4.18: the boundary relations on the triangulation Δ correspond to the incidence relations on the dual cell decomposition D. The difficult part of Poincaré duality is getting the signs correct in the diagram

$$
\begin{CD}
\cdots @>>> C^{k-1} @>d^{k-1}>> C^k @>>> \cdots \\
@. @VV\phi_{k-1}V @VV\phi_k V @. \\
\cdots @>>> D_{n-k+1} @>d_{n-k+1}>> D_{n-k} @>>> \cdots
\end{CD}
$$

The construction of ϕ runs as follows. First, orient the top dimensional simplices of Δ so that

$$\omega = \sum_{\sigma_i \in \Delta_n} \sigma_i$$

satisfies $\partial_n(\omega) = 0$. This is possible since Δ triangulates a manifold M, and $\partial(M) = 0$. A basis for $C_n(\Delta)$ is given by the $\sigma \in \Delta_n$, so a basis for $C^n(\Delta)$ is given by the $\sigma^\vee$. The dual block to σ is the zero-cell σ^* corresponding to $b(\sigma)$, so we have a map $\phi_n : C^n \to D_0$ which sends $\sigma^\vee$ to σ^*. To define ϕ_{n-1}, let ξ be an oriented $n - 1$ face of Δ. We require

$$d_1 \cdot \phi_{n-1}(\xi^\vee) = \phi_n d^{n-1}(\xi^\vee).$$

There exist a pair of faces $\sigma_i \in \Delta_n$ so that $\xi = \sigma_1 \cap \sigma_2$. Since $\partial_n(\omega) = 0$, ξ appears with opposite signs in the expressions for $\partial(\sigma_1)$ and $\partial(\sigma_2)$, so we can assume

$$d^{n-1}(\xi^\vee) = \sigma_1^\vee - \sigma_2^\vee, \text{ which implies } \phi_n \cdot d^{n-1}(\xi^\vee) = \sigma_1^* - \sigma_2^*.$$

Define $\phi_{n-1}(\xi^\vee) = [\sigma_2^*, \xi^*] + [\xi^*, \sigma_1^*]$. Then we have

$$d_1(\phi_{n-1}(\xi^\vee)) = d_1([\sigma_2^*, \xi^*]) + d_1([\xi^*, \sigma_1^*]) = \sigma_1^* - \sigma_2^*, \text{ as required.}$$

The general recipe for ϕ is as follows: let τ be an element of C_k, and $\tau^\vee$ the corresponding element of C^k. Then the corresponding element of D_{n-k} is obtained as

$$\phi_k(\tau^\vee) = \sum_{\substack{\sigma \in \Delta_n \\ \tau \in \Delta_k \\ \tau \subset \sigma}} \text{sgn}(\tau, \sigma)\tau_\sigma, \text{ where } \text{sgn}(\tau, \sigma) \text{ is the sign of } \tau \in \partial(\sigma).$$

A computation shows that

$$\phi_k \cdot d^{k-1} = d_{n-k+1} \cdot \phi_{k-1}.$$

This concludes our sketch of the proof of Poincaré duality.   □

When M is compact but not orientable, Theorem 5.4.1 holds, with $\mathbb{Z}/2$ coefficients. In this case, we don't need to worry about ϕ, because $1 = -1$, which simplifies the proof.

Example 5.4.19 The differential for the Klein bottle was computed in Example 5.3.4. Let σ be the two-cell, and τ_1, τ_2 the vertical and horizontal edges.

$$\partial_2(\sigma) = 2 \cdot \tau_2,$$

With $\mathbb{Z}/2$ coefficients, we see that the homology of the Klein bottle is the same as that of the torus, so

$$\mathbf{b}_{\mathbb{Z}/2}(\text{Klein Bottle}) = (1, 2, 1)$$

and Poincaré duality holds.

In [45], Cohen-Steiner–Edelsbrunner–Harer extend the birth-death pairings of persistent homology (Chap. 7) to pairings involving persistent classes (which have a birth, but not a death), and prove a Poincaré duality theorem in that context.

Chapter 6
Persistent Algebra: Modules Over a PID

A fundamental result in topological data analysis is that persistent homology can be represented as a barcode; short bars in the barcode represent homology classes that are transient whereas long bars represent persistent features.

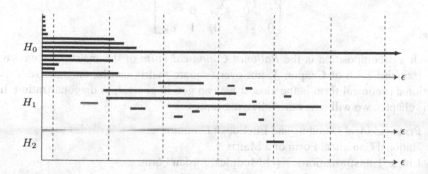

The intuition is that noise in the data corresponds to features that have short bars; long bars represent underlying structure in the data. The result follows from a central structure theorem in algebra: the decomposition theorem for modules over a principal ideal domain. This chapter focuses on the structure theorem; one surprise is that the theorem has applications ranging from the theory of finitely generated Abelian groups to the Jordan block decomposition of matrices.

Theorem 6.0.1 *A finitely generated module M over a principal ideal domain (PID) R has a direct sum decomposition as in Definition 2.1.5:*

$$M \simeq \left(\bigoplus_{i=1}^{m} R/\langle d_i \rangle \right) \bigoplus R^n \text{ with } d_i \mid d_{i+1} \text{ and } 0 \neq d_i \text{ a nonunit.}$$

© The Author(s), under exclusive license to Springer Nature Switzerland AG 2022
H. Schenck, *Algebraic Foundations for Applied Topology and Data Analysis*,
Mathematics of Data 1, https://doi.org/10.1007/978-3-031-06664-1_6

Since R is a principal ideal domain, $d_i | d_{i+1}$ is equivalent to $d_{i+1} \in \langle d_i \rangle$. The term R^n is the *free* component of M, and the term $\oplus R/\langle d_i \rangle$ is the *torsion* component. Recall that for an R-module M and $m \in M$, $\operatorname{ann}(m) = \{r \in R \mid rm = 0\}$. If R is a domain, an R-module M is *torsion* if $\operatorname{ann}(m) \neq 0$ for all $m \in M$. For example $R/\langle d_i \rangle$ is torsion. In the context of persistent homology, the free component corresponds to long bars, and the torsion component corresponds to short bars. In linear algebra, a consequence of Theorem 6.0.1 is that a linear operator

$$V \xrightarrow{T} V$$

on a finite dimensional vector space V can be written after a change of basis as a block diagonal matrix, where the blocks along the diagonal are of the form

$$\begin{bmatrix} 0 & 0 & \cdots & 0 & a_1 \\ 1 & \ddots & \cdots & 0 & a_2 \\ 0 & 1 & \ddots & 0 & \vdots \\ \vdots & 0 & \ddots & 0 & \vdots \\ 0 & \cdots & 0 & 1 & a_m \end{bmatrix} \tag{6.0.1}$$

Such a decomposition is the *Rational Canonical Form* of the matrix. As we saw in Exercise 1.3.6 of Chapter 1, not every square matrix can be diagonalized; the rational canonical form is the closest we can get, in general, to diagonalization. In this chapter, we will cover the following topics:

- Principal Ideal Domains and Euclidean Domains.
- Rational Canonical Form of a Matrix.
- Linear Transformations, $\mathbb{K}[t]$-Modules, Jordan Form.
- Structure of Abelian Groups and Persistent Homology.

6.1   Principal Ideal Domains and Euclidean Domains

Exercise 2.2.13 of Chap. 2 defined a principal ideal domain: a ring which is an integral domain ($a \cdot b = 0$ implies either a or b is zero), and where every ideal is principal (so can be generated by a single element). In §4 we prove Theorem 6.0.1 with the additional hypothesis that the PID is a Euclidean Domain.

Definition 6.1.1 An integral domain R is a *Euclidean Domain* if it possesses a Euclidean norm, which is a function

$$R \setminus 0 \xrightarrow{v} \mathbb{Z}_{\geq 0} \text{ such that}$$

- For all $a, b \in R$ with $b \neq 0$, there exist $q, r \in R$ such that

$$a = bq + r, \text{ with } r = 0 \text{ or } v(r) < v(a).$$

- For all $a \neq 0 \neq b \in R$, $v(a) \leq v(a \cdot b)$

Theorem 6.1.2 *A Euclidean Domain R is a Principal Ideal Domain.*

Proof Let $I \subseteq R$ be an ideal; if $I = \langle 0 \rangle$ there is nothing to prove, so suppose I is nonzero. Choose $f \in I$ with $v(f)$ minimal. Any $g \in I$ satisfies

$$g = f \cdot q + r,$$

with $r = 0$ or $v(r) < v(f)$. Since

$$r = (g - f \cdot q) \in I$$

and f was chosen with $v(f)$ minimal, $v(r)$ cannot be less than $v(f)$, so $r = 0$, hence any g is a multiple of f and $I = \langle f \rangle$.   □

Example 6.1.3 Examples of Euclidean norms

- $R = \mathbb{Z}$, with norm $v(a) = |a|$.
- $R = \mathbb{K}[x]$, with norm $v(f(x)) = \text{degree}(f(x))$.

Exercise 6.1.4 Prove the two norms appearing in Example 6.1.3 are Euclidean norms, so $\mathbb{Z}$ and $\mathbb{K}[x]$ are Euclidean Domains, and hence by Theorem 6.1.2 also Principal Ideal Domains.   ◊

Algorithm 6.1.5 *The Euclidean Algorithm. In a Euclidean domain, the Euclidean algorithm allows us to find the greatest common divisor (GCD) of two nonzero elements: the largest c such that $c \mid a$ and $c \mid b$. To compute $\text{GCD}(a, b)$*

(a) Form $a = bq_1 + r_1$, where $r_1 = 0$ or $v(r_1) < v(b)$.
(b) If $r_1 = 0$ then $b|a$, done. Else since $v(r_1) < v(b)$, $b = r_1 q_2 + r_2$.
(c) Iterate until $r_{i-1} = r_i q_{i+1} + r_{i+1}$ with $r_{i+1} = 0$

The algorithm terminates, since the r_i are a decreasing sequence in $\mathbb{Z}_{\geq 0}$, and

$$\text{GCD}(a, b) = \text{ the last nonzero } r_i$$

which follows by reversing the process.

Example 6.1.6 To compute $\text{GCD}(48, 332)$, we proceed as follows:

(a) $332 = 48 \cdot 6 + 44$.
(b) $48 = 44 \cdot 1 + 4$
(c) $44 = 4 \cdot 11 + 0$

So $\text{GCD}(48, 332) = 4$. To check, note that $48 = 4 \cdot 12$ and $332 = 4 \cdot 83$.

Exercise 6.1.7 GCD computations.

(a) Show $\mathrm{GCD}(4, 83) = 1$.
(b) Show $\mathrm{GCD}(462, 90) = 6$.
(c) Show $\mathrm{GCD}(105, 126) = 21$.

It is nontrivial to construct PID's which are not Euclidean Domains.   ◇

6.2   Rational Canonical Form of a Matrix

As noted in the introduction to this chapter, the Rational Canonical Form (RCF) of a square matrix is, in general, the closest we can get to diagonalization for an arbitrary square matrix. Recall from Definition 1.3.3 that square matrices A and B are similar if there is an invertible C such that $CAC^{-1} = B$.

Definition 6.2.1 An invariant block is an $m \times m$ matrix of the form of Eq. 6.0.1: zero except in positions $a_{i,i-1} = 1$ and $a_{i,m}$.

Theorem 6.2.2 *Any $n \times n$ matrix is similar to a block diagonal matrix*

$$\begin{bmatrix} B_{i_1} & 0 & \cdots & \cdots & 0 \\ 0 & B_{i_2} & \ddots & \cdots & 0 \\ \vdots & 0 & \ddots & 0 & \vdots \\ 0 & 0 & \ddots & 0 & \vdots \\ 0 & \cdots & \cdots & 0 & B_{i_k} \end{bmatrix} \tag{6.2.1}$$

with B_{i_j} an invariant block of size i_j, and

$$\sum_{j=1}^{k} i_j = n.$$

Proof An $n \times n$ matrix represents, with respect to some basis, a linear transformation $T : V \to V$. Take an element $v \in V$, and let j be the smallest integer such that

$$T^j(v) \in \mathrm{Span}(B), \text{ with } B = \{v, T(v), T^2(v), \cdots, T^{j-1}(v)\}.$$

If we write V_{j-1} for $\mathrm{Span}(B)$, then B is a basis for V_{j-1}: by construction it is a spanning set, and linearly independent by the choice of j as minimal. Note that T restricts to a linear transformation on V_{j-1}. Write

$$T^j(v) = \sum_{i=0}^{j-1} a_i T^i(v).$$

As $v \mapsto T(v)$, $T(v) \mapsto T^2(v)$ and so on, with respect to the ordered basis B, the matrix for T restricted to V_{j-1} is

$$
\begin{bmatrix}
0 & 0 & \cdots & 0 & a_0 \\
1 & \ddots & \cdots & 0 & a_1 \\
0 & 1 & \ddots & 0 & \vdots \\
\vdots & 0 & \ddots & 0 & \vdots \\
0 & \cdots & 0 & 1 & a_{j-1}
\end{bmatrix}
\tag{6.2.2}
$$

Now iterate the process for V/V_{j-1}. □

Exercise 6.2.3 Find the RCF of the 4×4 matrix in Chap. 1, Example 1.4.7. ◇

6.3 Linear Transformations, $\mathbb{K}[t]$-Modules, Jordan Form

In this section, we connect the rational canonical form of a matrix to finitely generated modules over the ring $\mathbb{K}[t]$.

Theorem 6.3.1 *There is a one to one correspondence between the set of linear transfomations $V \to V$ of a finite dimensional vector space V over $\mathbb{K}$, and the set of finitely generated torsion $\mathbb{K}[t]$-modules.*

Proof First, lct

$$V \xrightarrow{T} V$$

be a linear transformation, and $f(t) \in \mathbb{K}[t]$ be given by

$$f(t) = \sum_{i=0}^{m} a_i t^i.$$

Define an action of $f(t)$ on $v \in V$ via

$$f(t) \cdot v = \sum_{i=0}^{m} a_i T^i(v).$$

A check shows that this satisfies the module axioms of Chap. 2, Definition 2.2.1. Now suppose M is a finitely generated torsion $\mathbb{K}[t]$-module, so the constants $\mathbb{K}$ act on M, and the module properties of this action are identical to the vector space properties, so M is a vector space over $\mathbb{K}$. To determine the linear transformation T, consider the action of multiplication by t. The module properties imply that

$$t \cdot (c\mathbf{v} + \mathbf{w}) = ct \cdot \mathbf{v} + t \cdot \mathbf{w} \text{ for } c \in \mathbb{K} \text{ and } \mathbf{v}, \mathbf{w} \in M.$$

So multiplication by the element t yields a linear transformation $T(\mathbf{v}) = t \cdot \mathbf{v}$. □

Example 6.3.2 Suppose $V = \mathbb{K}[t]/\langle f(t)\rangle$, where $f(t) = t^j + a_{j-1}t^{j-1} + \cdots + a_0$. Then with respect to the basis

$$\{1, t, t^2, \ldots, t^{j-1}\}$$

the linear transformation T above is represented by the matrix of Eq. 6.2.2, with a_i replaced by $-a_i$.

Example 6.3.3 For an algebraically closed field such as $\mathbb{K} = \mathbb{C}$, every polynomial in $\mathbb{K}[t]$ factors as a product of linear forms. The *Jordan Canonical Form* of a matrix exploits this fact; albeit with a slightly different basis. First, let

$$f(t) = (t - \alpha)^n \text{ and } W = \mathbb{K}[t]/\langle f(t)\rangle.$$

Let w_0 denote the class of 1 in W, and $w_i = (t - \alpha)^i \cdot w_0$. Then

$$
\begin{array}{ccc}
(t - \alpha) \cdot w_0 & = & w_1 \\
\vdots & \vdots & \vdots \\
(t - \alpha) \cdot w_{n-2} & = & w_{n-1} \\
(t - \alpha) \cdot w_{n-1} & = & 0.
\end{array}
$$

Letting $T = \cdot t$, we have that

$$T(w_i) = \alpha w_i + w_{i+1} \text{ for } i \in \{0, \ldots, n-2\}$$

and $T(w_{n-1}) = \alpha w_{n-1}$. In this basis, the matrix for T is

$$
T = \begin{bmatrix}
\alpha & 0 & \cdots & \cdots & 0 \\
1 & \alpha & 0 & \ddots & \vdots \\
0 & 1 & \ddots & \ddots & \vdots \\
\vdots & \ddots & \ddots & \alpha & 0 \\
0 & \cdots & 0 & 1 & \alpha
\end{bmatrix}.
$$

Reordering the basis so the 1's are above the diagonal gives a matrix which is called a *Jordan block*. In general, over an algebraically closed field we may factor $f(t) = \prod_{j=1}^{k}(t - \alpha_j)^{m_j}$, and in the representation of Theorem 6.2.2 the blocks B_{i_j} are replaced with Jordan blocks with α_j on the diagonal.

6.4 Structure of Abelian Groups and Persistent Homology

Our second application, and the one which is of key importance in topological data analysis, comes from finitely generated Abelian groups. First, note that an Abelian group A is a $\mathbb{Z}$-module, because for $n \in \mathbb{Z}_{>0}$ and $a \in A$,

$$n \cdot a = a + a + \cdots + a \text{ for } n\text{- copies of } a.$$

In this case, Theorem 6.0.1 may be stated in the form

Theorem 6.4.1 *A finitely generated Abelian group has a direct sum decomposition*

$$A \simeq \mathbb{Z}^n \bigoplus \mathbb{Z}/\langle p_i^{\alpha_i} \rangle, \text{ with } p_i \text{ prime.}$$

The proof of Theorem 6.4.1 works for any Euclidean Domain, for our data analysis application of persistent homology we will replace $\mathbb{Z}$ with $\mathbb{K}[x]$.

Theorem 6.4.2 *For A an $m \times n$ matrix with entries in $\mathbb{Z}$, there exist integer matrices B and C which encode elementary row and column operations over $\mathbb{Z}$, such that*

$$A' = BAC^{-1} = \begin{bmatrix} d_1 & 0 & \cdots & \cdots & \cdots & 0 \\ 0 & d_2 & 0 & \cdots & \cdots & 0 \\ \ddots & 0 & \ddots & 0 & \cdots & 0 \\ \vdots & & \ddots & d_k & 0 & \vdots \\ 0 & 0 & \ddots & \ddots & 0 & \vdots \\ 0 & \cdots & \ddots & \ddots & \cdots & 0 \end{bmatrix} \text{ with } d_i > 0 \text{ and } d_i | d_{i+1}.$$

Proof The proof has two steps, and moves back and forth between them:

(a) We first use elementary row and column swaps to put the smallest entry of A in position $(1, 1)$. Next, use elementary row and column operations to zero out all entries in the first row and column, except at position $(1, 1)$. When the entries of A are in a field $\mathbb{K}$, this is straightforward, but over $\mathbb{Z}$ care must be taken. For example, if the first row starts as $[2, 3, \cdots]$, we can use 2 to reduce 3, with a remainder of 1. Now 1 is the smallest entry, so swap it into position $(1, 1)$, and continue.

The key point is that each entry in the first row may be written as

$$a_{1i} = a_{11}q + r, \text{ with } r = 0 \text{ or } r < a_{11}.$$

If $r = 0$ then a_{1i} reduces to zero mod a_{11}. Otherwise after reducing $a_{1i} = r$, swap the first column and column i to get the smaller value r in position $(1, 1)$. This process concludes with a matrix A' which is nonzero in the first row and first column only in position $(1, 1)$, with $A'_{1,1} = d_1$. To achieve $d_i | d_{i+1}$, we need to do more work.

(b) Suppose A' has an entry b which is not divisible by d_1. Add the column of A' containing b to the first column of A', and call the resulting matrix A''. Because the column of A' containing b has zero entry in the first row, and the first column of A' is zero except in position $(1, 1)$, the first column of A'' has b as an entry. Since $a_{11} = d_1$ does not divide b, we have

$$b = a_{11}q + r \text{ with } r \neq 0, \text{ so } r < a_{11}.$$

Swap the row with r with the first row, and return to step (a).

The process terminates: at every step, the numbers decrease in absolute value.   □

Exercise 6.4.3 Working with $\mathbb{Z}/p$.

(a) Show that

$$\mathbb{Z}/2 \times \mathbb{Z}/2 \not\simeq \mathbb{Z}/4.$$

(b) Show that if p_1 and p_2 are distinct prime numbers, then

$$\mathbb{Z}/p_1 \times \mathbb{Z}/p_2 \simeq \mathbb{Z}/p_1 p_2.$$

Now show that Theorem 6.4.1 is equivalent to Theorem 6.4.2.   ◇

Lemma 6.4.4 *If R is a Noetherian ring and M a finitely generated R-module, then the module of relations F in Definition 2.2.20 of Chap. 2 is of finite rank.*

Proof Exercise 2.4.4 of Chap. 2 shows that a finitely generated module over a Noetherian ring is Noetherian, hence so is every submodule.   □

With these preliminaries in hand, we're ready for the proof of Theorem 6.4.1.

Proof (of Theorem 6.4.1) If A is a finitely generated Abelian group, then A is a finitely generated $\mathbb{Z}$-module. As a principal ideal domain is Noetherian, Lemma 6.4.4 implies that A has a presentation of the form

$$\mathbb{Z}^m \xrightarrow{\phi} \mathbb{Z}^n \longrightarrow A \longrightarrow 0.$$

By Theorem 6.4.2,

$$\phi = \begin{bmatrix} d_1 & 0 & \cdots & & \cdots & & 0 \\ 0 & d_2 & 0 & \cdots & & \cdots & 0 \\ & \ddots & 0 & \ddots & 0 & \cdots & 0 \\ \vdots & & \ddots & \ddots & d_k & 0 & \vdots \\ 0 & 0 & & \ddots & \ddots & 0 & \vdots \\ 0 & \cdots & & \ddots & \ddots & \cdots & 0 \end{bmatrix} \text{ with } d_i > 0 \text{ and } d_i | d_{i+1}.$$

Since $A = \mathrm{coker}(\phi)$, this means exactly that $A \simeq \left(\bigoplus_{i=1}^{k} \mathbb{Z}/d_i \mathbb{Z} \right) \oplus \mathbb{Z}^{n-k}$.

For a Euclidean domain, Theorems 6.4.1 and 6.0.1 are equivalent, and with small modifications this proof also works for a PID. □

Exercise 6.4.5 Row reduce the matrix

$$\begin{bmatrix} 3 & 1 & -4 \\ 2 & -3 & 1 \\ -4 & 6 & -2 \end{bmatrix}$$

using integer elementary operations. ◇

6.4.1 $\mathbb{Z}$-Graded Rings

Chapter 7 introduces *persistent homology*, where the algebraic invariants that are constructed are modules over $\mathbb{K}[x]$. Since $\mathbb{K}[x]$ is a Euclidean Domain, we can apply Theorem 6.4.1 to analyze the structure of the resulting modules. In fact, $\mathbb{K}[x]$ has additional structure: it is a $\mathbb{Z}$-graded ring:

Definition 6.4.6 A ring R is $\mathbb{Z}$-graded if $R = \oplus_{i \in \mathbb{Z}} R_i$, such that if $r_i \in R_i$ and $r_j \in R_j, r_i \cdot r_j \in R_{i+j}$. Note that if $R_0 = \mathbb{K}$ then all the R_i are $\mathbb{K}$-vector spaces. A $\mathbb{Z}$-graded module M over a $\mathbb{Z}$-graded ring R must satisfy a similar condition

$$M = \bigoplus_{i \in \mathbb{Z}} M_i, \text{ and if } m \in M_i \text{ and } r_j \in R_j \text{ then } r_j \cdot m_i \in M_{i+j}.$$

We write $R(i)$ for a free module of rank one, with generator in degree $-i$. Therefore $R(i)_j \simeq R_{i+j}$, in particular, $R(-i)_i \simeq R_0$.

Exercise 6.4.7 For $R = \mathbb{K}[x]$, $R(-2)$ is a free module with unit appearing in degree 2, and so multiplication by x^2 defines a graded map $R \to R(2)$. Show that for $R = \mathbb{K}[x_1, \ldots, x_n]$ the multiplication by $f(x_1, \ldots, x_n)$ defines a graded map only when f is a homogeneous polynomial as in Exercise 2.4.16. ◇

Grading plays a key role in persistent homology, and is discussed in detail in Sect. 8.2.

Example 6.4.8 We return to the figure at the beginning of this chapter. As noted in the introduction, the free component in the decomposition is graphically represented by long bars in the barcode, and the torsion component is represented by short bars. The graded condition means that a free summand will correspond to a *graded* ideal in $R = \mathbb{K}[x]$. Since we know that $\mathbb{K}[x]$ is a principal ideal domain, any ideal in $\mathbb{K}[x]$ has the form $\langle f(x) \rangle$.

In Exercise 6.4.9 below, you'll show that the graded condition means that $f(x)$ must be a monomial, hence of the form x^i. The *birth* of a long bar at time i corresponds exactly to $\langle x^i \rangle$. Similarly, a graded torsion summand corresponds to a graded ideal $\langle x^j \rangle$. Continuing to think of the exponent as representing time, $\langle x^j \rangle$ represents a birth at time j; because the class is torsion it *dies* at a later time $\langle x^{j+k} \rangle$. To summarize, we have

$$
\begin{aligned}
\text{long bar} &\leftrightarrow & \langle x^{b_i} \rangle &\leftrightarrow & R(-b_i) \\
\text{short bar} &\leftrightarrow & \langle x^{b_i} \rangle / \langle x^{b_i + d_i} \rangle &\leftrightarrow & R(-b_i)/\langle x^{b_i+d_i} \rangle
\end{aligned}
$$

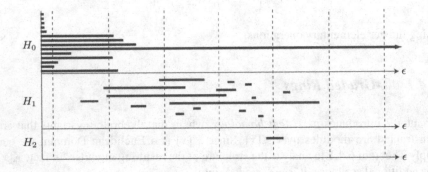

For example, H_0 has 13 short bars, and one long bar. The corresponding decomposition is:

$$
H_0 = \left(\bigoplus_{i=1}^{13} R/\langle x^{d_i} \rangle \right) \bigoplus R^1.
$$

$$
H_1 = \bigoplus_{j=1}^{15} \langle x^{b_j} \rangle / \langle x^{b_j + d_j} \rangle.
$$

$$
H_2 = \langle x^{b_j} \rangle / \langle x^{b_j + d_j} \rangle.
$$

In particular, note that for H_0 all classes are born at time zero, so the generators correspond to $\langle x^0 \rangle = R$. Similarly, for H_1 and H_2, all classes are torsion, so will have both a birth time b_j and a death time $b_j + d_j$.

Exercise 6.4.9 Prove that a proper nonzero $\mathbb{Z}$-graded ideal in $\mathbb{K}[x]$ must be of the form $\langle x^i \rangle$ for some $i \in \mathbb{Z}_{>0}$.   ◇

Exercise 6.4.10 For an R-module M, the annihilator is defined as

$$
\mathrm{Ann}(M) = \{ r \in R \mid r \cdot m = 0 \text{ for all } m \in M \}.
$$

Show $\mathrm{Ann}(M)$ is an ideal. What is the annihilator of $\bigoplus_{j=1}^k \langle x^{b_j} \rangle / \langle x^{b_j + d_j} \rangle$?   ◇

Chapter 7
Persistent Homology

We're ready to reap the fruits of our labors in the previous chapters, and attack the motivating problem from the preface: given point cloud data $X \subseteq \mathbb{R}^n$, we want to extract meaning from X. In this chapter, we study *persistent homology* (PH), which can facilitate detection and analysis of underlying structure in large datasets. Persistent homology assigns a graded module over a principal ideal domain to a filtered simplicial complex. In Chap. 6, we proved that a finitely generated module M over a principal ideal domain R has a decomposition:

$$ M \simeq (\bigoplus_{i=1}^{k} R/p_i) \bigoplus R^m $$

Persistent homology is often displayed as a barcode diagram. The decomposition above translates into a barcode as follows: the free summands (which are known as *long bars*) represent *persistent* features of the underlying data set, while the torsion components (which are known as *short bars*) represent noise in the data. The grading allows us to assign *birth* and *death* times to homology classes. Persistent homology has proven effective in a number of settings, including breast cancer pathology [28], viral evolution [37], and visual cortex activity [136], to name just a few. We start by using X as a seed from which to build a family of spaces

$$ X_\epsilon = \bigcup_{p \in X} N_\epsilon(p), \text{ where } N_\epsilon(p) \text{ denotes an } \epsilon \text{ ball around } p. $$

As $X_\epsilon \subseteq X_{\epsilon'}$ if $\epsilon \leq \epsilon'$, the result is a filtered topological space. This chapter covers

- Barcodes, Persistence Diagrams, Bottleneck Distance.
- Morse Theory.
- The Stability Theorem.
- Interleaving and Categories.

© The Author(s), under exclusive license to Springer Nature Switzerland AG 2022
H. Schenck, *Algebraic Foundations for Applied Topology and Data Analysis*,
Mathematics of Data 1, https://doi.org/10.1007/978-3-031-06664-1_7

7.1   Barcodes, Persistence Diagrams, Bottleneck Distance

In this section the focus of our interest will be the Čech or Rips complex obtained from point cloud data input. Recall from §4.4 that these are filtered simplicial complexes, varying with a parameter $\epsilon \in \mathbb{R}$.

Example 7.1.1 Consider the filtered simplicial complex below, taken from [78]. The ranks of the homology are depicted as barcodes. There are 7 "snapshots" of the corresponding simplicial complex Δ_ϵ.

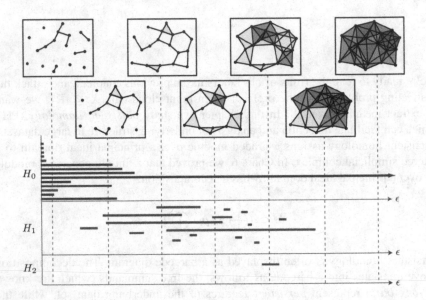

For example, at the time the 3rd snapshot is taken, the simplicial complex Δ_ϵ has a single connected component, which is reflected in the fact that for the corresponding value of ϵ, there is a single long bar for H_0. Similarly, at the time of the sixth snapshot, all the one and two dimensional holes in Δ_ϵ have been filled in, which is reflected that there are no classes surviving (no *bars*) for H_1 and H_2.

7.1.1   History

The idea of extracting meaning from points sampled from a space reaches back to the dawn of topology–at least to [117], and in various forms even earlier. A major hurdle was the difficulty (or perhaps, impossibility) of effective implementation; this became surmountable with the advent of high speed computing. The modern development of the theory has roots in work of Frosini [73], [74] and Frosini–Landi [75] on connected components in the early 1990's, and was extended to higher homology by Robins [130] and Edelsbrunner–Letscher–Zomorodian [65]. From there the theory of persistent homology took off in leaps and bounds; a nice synopsis of the story up to 2008 appears in the survey [66]. Carlsson–Zomorodian introduced multiparameter persistent homology (MPH) in [33], and we tackle MPH in the next chapter. A detailed treatment of stability (Sect. 7.3) appears in the monograph [40] of Chazal–de Silva–Glisse–Oudot.

7.1.2 Persistent Homology and the Barcode

As discussed in the preface, the starting point for persistent homology is a point cloud dataset X, from which we build a filtered topological space

$$X_\epsilon = \bigcup_{p \in X} N_\epsilon(p), \text{ where } N_\epsilon(p) \text{ denotes an } \epsilon \text{ ball around } p.$$

The salient feature is that as in Morse theory (which we tackle in the next section), for both the Čech and Rips complexes the topology of the spaces X_ϵ and Δ_ϵ only changes at a finite number of values of ϵ. For $\epsilon < \epsilon'$, the inclusions

$$X_\epsilon \hookrightarrow X_{\epsilon'} \text{ and } \Delta_\epsilon \hookrightarrow \Delta_{\epsilon'}$$

induce maps in homology.

Definition 7.1.2 The pth persistent homology is the image of the homomorphism

$$H_p(\Delta_\epsilon) \xrightarrow{f_p^{\epsilon,\epsilon'}} H_p(\Delta_{\epsilon'})$$

When $\epsilon' - \epsilon$ is small, these maps are uninteresting, save at the finite number of parameter values where a new simplex is added in. The finiteness aspect means that the problem can be discretized (see, for example, Carlsson–Zomorodian [32]): reindex Δ using $\mathbb{Z}$ as an index set, with an inclusion

$$\Delta(i) \hookrightarrow \Delta(i+1)$$

when a new simplex is added. For a fixed index i, there is the usual simplicial boundary map on the chain complex $C_\bullet(\Delta(i))$. Using $R = \mathbb{K}[x]$ with $\mathbb{K}$ a field as the coefficient ring allows us to encode the passage from $\Delta(i)$ to $\Delta(i+1)$ by multiplication by x. This yields a double complex of *graded modules*, which are discussed in detail in Chap. 8. For a fixed value i

$$\cdots \longrightarrow C_{j+1}(\Delta(i), R) \xrightarrow{\partial_{j+1}} C_j(\Delta(i), R) \longrightarrow \cdots$$

is a complex of modules over a principal ideal domain, and the structure theorem of Chap. 6 gives a complete description of the homology.

Definition 7.1.3 A homology class α that first appears in $\Delta(i)$ is said to be *born* at time i. If it does not vanish at a later time then it *persists*, and corresponds to the free module $x^i \cdot R$, depicted as a *long bar* as in Example 7.1.1. If α first vanishes (*dies*) in $\Delta(i+j)$ then the corresponding module is $x^i \cdot R/\langle x^{i+j}\rangle$, depicted in Example 7.1.1 as a *short bar*. The *barcode* of a point cloud data set X is a multiset of bars (several classes can have coeval births and deaths), yielding a graphic representation of the PH of a filtered simplicial complex. It is a *complete invariant*: the homology is equivalent to the barcode, as we saw in Example 6.4.8.

7.1.3   Computation of Persistent Homology

In [65], Edelsbrunner-Letscher-Zomorodian give an algorithm (often called the *standard algorithm*) for computing persistent homology. Large data sets present considerable computational hurdles, and there have been many papers on extending or improving the algorithm. See for example [32], [97], [105] and references therein, and [123] for an overview.

The main idea is to set up a pairing where birth-death (or birth) classes in homology have a distinguished representative. This echoes Morse theory, where critical points carry a homological signal. The first step is to package all the boundary maps ∂_i into a single large matrix D. Let $\Delta(i)$ be the ith term in the filtration, and let S be the set of all simplices (of all dimensions) appearing in the filtration, with $|S| = n$. We define $n \times n$ matrices D and R as follows:

(a) Linearly order S, such that $\sigma_i \prec \sigma_j$ for $i \leq j$ when either $\sigma_i \subseteq \sigma_j$ or $\sigma_j \subseteq \Delta(j) \setminus \Delta(i)$ and $\sigma_i \in \Delta(i)$ (so consistent with order of appearance).

(b) $D_{ij} = \mathrm{sgn}(\sigma_i, \sigma_j)$ if σ_i is a codimension one face of σ_j, else $D_{ij} = 0$.

(c) For $j \in \{1, \ldots, n\}$, define $\mathrm{low}(j)$ to be the largest row index of a nonzero entry (i.e. the lowest or bottom-most) in column j; if column j contains all zeros, set $\mathrm{low}(j) = 0$. A matrix is in *reduced* form if for every column j having $\mathrm{low}(j) = i \neq 0$, there is only one nonzero entry in row i.

(d) Do left to right column operations to transform D to a reduced matrix R.

The reduced matrix R encodes PH: if $\mathrm{low}(j) = i$, and $\dim(\sigma_i) = m$, then σ_i signals birth of a class in H_m, and σ_j the demise. If $\mathrm{low}(j) = 0$ and $j \neq \mathrm{low}(k)$ for any k, then there is a birth at σ_j. An example is worth a thousand words:

Example 7.1.4 Consider the complex Δ of Exercise 2.2.8, filtered as below

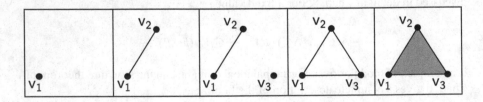

The order $\{[1], [2], [12], [3], [13], [23], [123]\}$ satisfies (a) above, yielding

$$D = \begin{bmatrix} 0 & 0 & 1 & 0 & 1 & 0 & 0 \\ 0 & 0 & -1 & 0 & 0 & 1 & 0 \\ 0 & 0 & 0 & 0 & 0 & 0 & 1 \\ 0 & 0 & 0 & 0 & -1 & -1 & 0 \\ 0 & 0 & 0 & 0 & 0 & 0 & -1 \\ 0 & 0 & 0 & 0 & 0 & 0 & 1 \\ 0 & 0 & 0 & 0 & 0 & 0 & 0 \end{bmatrix} \text{ and } R = \begin{bmatrix} 0 & 0 & 1 & 0 & 1 & 0 & 0 \\ 0 & 0 & -1 & 0 & 0 & 0 & 0 \\ 0 & 0 & 0 & 0 & 0 & 0 & 1 \\ 0 & 0 & 0 & 0 & -1 & 0 & 0 \\ 0 & 0 & 0 & 0 & 0 & 0 & -1 \\ 0 & 0 & 0 & 0 & 0 & 0 & 1 \\ 0 & 0 & 0 & 0 & 0 & 0 & 0 \end{bmatrix}$$

We obtain classes in PH as below. The class corresponding to [1] is persistent.

$\text{low}([12]) = [2]$. [2] born at time 1, dies at time 1 $\rightarrow$ no class.
$\text{low}([13]) = [3]$. [3] born at time 2, dies at time 3 $\rightarrow$ class in H_0.
$\text{low}([123]) = [23]$. [23] born at time 3, dies at time 4 $\rightarrow$ class in H_1.
$\text{low}([1]) = 0$. [1] born at time 0, $\nexists k \mid \text{low}(k) = [1]$ $\rightarrow$ class in H_0.

7.1.4 Alpha and Witness Complexes

As noted in Chap. 4, the Čech complex requires determination of the intersections between neighborhoods of points, which can be computationally intensive. A first analysis of the runtime for the reduction algorithm in the previous section might indicate that it has complexity $\mathcal{O}(n^3)$, but [112] shows this can be improved to $\mathcal{O}(n^{2.376})$.

One way to improve runtime is to construct variants of the Čech complex which involve a smaller number of simplices. Below we give a quick overview of two other commonly used complexes, the *Alpha* complex introduced by Edelsbrunner-Kirkpatrick-Seidel in [64] and the *Witness* complex introduced by de Silva–Carlsson in [51]. To describe the Alpha complex, we need to define Voronoi cells:

Definition 7.1.5 For a finite point set $X \subseteq \mathbb{R}^d$, the Voronoi cell of $x \in X$ is

$$V(x) = \{p \in \mathbb{R}^n \text{ such that } ||x - p|| \leq ||y - p|| \text{ for all } y \in X\}$$

For a parameter δ, the Alpha complex $A_\delta(X)$ is the nerve (as in Definition 4.1.5) of the cover consisting of the union of the sets $N_\delta(x) \cap V(x)$.

When $\delta \to \infty$, the intersections $N_\delta(x) \cap V(x)$ are just the Voronoi cells; the nerve of the resulting cover is the *Delaunay complex*. The Delaunay complex does not typically have a geometric realization, but a notable exception occurs when the points of X are sufficiently general (not too many points on a linear subspace or sphere), in which case the resulting geometric realization is a triangulation of the convex hull of X. The initial motivation came from the setting $X \subseteq \mathbb{R}^2$, where

using the Alpha complex results in a substantial speed up (coming from duality).
See [66] for a detailed discussion.

For the *Witness complex*, the key idea is to identify distinguished subsets of a finite
set X that encode the key features of the set.

Definition 7.1.6 Let $Y \subseteq X \subseteq \mathbb{R}^d$ with X finite. A point $p \in \mathbb{R}^d$ is a weak witness
at scale ϵ for Y with respect to X if $||p - y|| \leq ||p - x|| + \epsilon$ for all $y \in Y$,
$x \in X \setminus Y$. The witness complex $D(X, \epsilon)$ has a vertex for each $x \in X$; $[x_{i_0}, \ldots, x_{i_k}]$
is a k-simplex of $D(X, \epsilon)$ exactly if it has a weak witness at scale ϵ.

A point $p \in S$ is a weak witness for a subset Y of X when p is a sort of "center
of attraction" or "landmark" for the points of Y, while the points of $X \setminus Y$ remain
standoffish (at a further distance from p than the points of Y). Thus, in certain
circumstances, the cluster of points forming Y may represent a simplex of dimension
$\leq |Y| - 1$ in our original sample, so the set Y is in some sense being represented
by the single point p. The papers of Attali–Edelsbrunner–Mileyko [5] and Guibas–
Oudot [85] give conditions under which witness complexes can discern the topology
of curves and surfaces in $\mathbb{R}^d$. Both the Alpha complex and the Witness complex can
be generalized to other metric spaces.

7.1.5   Persistence Diagrams

The barcode is a graphical depiction of persistent homology, which highlights
visually which homology classes survive the longest. An equivalent presentation of
the information in a barcode is via a persistence diagram: a bar is plotted as a point
with coordinates (b, d), where b represents the appearance (birth) of a bar, and d
the death. It turns out that to set up a distance between barcodes the depiction of the
information as a persistence diagram is a better encoding. Chapter 7 of [67] gives a
detailed description of persistence diagrams and stability; points of the persistence
diagram are elements of the extended plane $\widetilde{\mathbb{R}^2}$, which we represent by adding a line
at infinity $\mathbb{R}^1_\infty$, as in the construction of the projective plane in Chap. 2.

Definition 7.1.7 A *persistence diagram* is a collection of points, each having an
integer multiplicity, in $\mathbb{R}^2 \cup \mathbb{R}^1_\infty$, obtained from persistent homology as follows.
The *multiplicity* $\mu_p^{i,j}$ is the number of classes in H_p which are born at time i and die
at time j; assign $\mu_p^{i,j}$ to the point (i, j). Similarly, $\mu_p^{i,\infty}$ is the number of classes in
H_p which are born at time i and persist; assign $\mu_p^{i,\infty}$ to the point $i \in \mathbb{R}^1_\infty$. Pictorially,
we superimpose $\mathbb{R}^1_\infty$ on $\mathbb{R}^2$ at a y value beyond the finite death times.

Thus, a point in the persistence diagram corresponds to a bar in the barcode, and the
multiplicity of the point is the number of copies of the bar. For $j \neq \infty$, the distance
from a point (i, j) in the persistence diagram to the diagonal line $x = y$ is $j - i$,
which is the length of the corresponding bar. Points of the form (i, ∞) correspond
to the long bars in the barcode which are born at time i.

Example 7.1.8 For the H_1 barcode of Example 7.1.1, the persistence diagram is below, with the horizontal time corresponding to the seven snapshots.

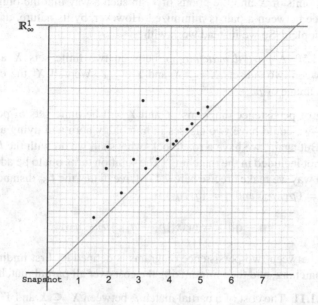

The longest H_1 bar is born just after the third snapshot is taken, and dies just before the sixth snapshot, and corresponds to the point $(3.1, 5.9)$. There are no persistent classes in H_1 for Example 7.1.1, hence no points on the line $\mathbb{R}^1_\infty$, and there are no simultaneous births/deaths, so all points off the diagonal have multiplicity one.

7.1.6 Metrics on Diagrams

For persistent homology to be useful, if two data sets X and Y are "close", then their persistence diagrams should be close. So our first task is to define a metric on persistence diagrams. The *Hausdorff metric* is a standard way to make the set of compact subsets of a metric space into a metric space itself.

Definition 7.1.9 Let X and Y be two non-empty subsets of a metric space with metric d. The Hausdorff distance between X and Y is

$$d_H(X, Y) = \max\{\sup_{x \in X} \inf_{y \in Y} d(x, y), \sup_{y \in Y} \inf_{x \in X} d(x, y)\}$$

Put simply, the Hausdorff distance is the longest distance between a point $x \in X$ and the point in Y closest to x.

To define a distance between sets X and Y corresponding to persistence diagrams, we take Hausdorff distance as a starting point. This entails setting up a matching between the points of X and the points of Y, in such a way that the biggest (think worst) distance between a pair is minimized. However, by its nature, data is often noisy or incomplete. So we instead work with

Definition 7.1.10 A *partial matching* μ between two finite sets X and Y is a bijection between two subsets $X' \subseteq X$ and $Y' \subseteq Y$. We call X' the coimage of μ, and Y' the image of μ.

In the setting of persistence diagrams, X and Y will be finite sets of points (with multiplicity—we could have two classes in homology born and dying at the same time) in $\mathbb{R}^2$. But again, we have a roadblock–what shall we do with the *leftovers*— those points not included in the matching? The solution turns out to be adding in the diagonal, in a way we make precise below. First, recall that the ℓ_∞ distance between two points $\mathbf{p} = (p_1, p_2)$ and $\mathbf{q} = (q_1, q_2)$ is

$$\|\mathbf{p} - \mathbf{q}\|_\infty = \max\{|p_1 - q_1|, |p_2 - q_2|\}. \qquad (7.1.1)$$

The distance between two persistence diagrams is defined by first finding the cost of a partial matching, and then minimizing the cost over all partial matchings:

Definition 7.1.11 The cost of a partial match μ between $X' \subseteq X$ and $Y' \subseteq Y$ is

$$
c(\mu) = \max\{ \sup_{\mathbf{p}' \in X'} \{\|\mathbf{p}' - \mu(\mathbf{p}')\|_\infty\},
$$
$$
\sup_{\mathbf{p} \in X \setminus X'} \{\tfrac{p_2 - p_1}{2}\}, \qquad (7.1.2)
$$
$$
\sup_{\mathbf{q} \in Y \setminus Y'} \{\tfrac{q_2 - q_1}{2}\}\}
$$

The *bottleneck distance* between two persistence diagrams X and Y is

$$d_B(X, Y) = \min\{c(\mu) \mid \mu \text{ a partial matching}\} \qquad (7.1.3)$$

Exercise 7.1.12 Show that for $\mathbf{p} = (p_1, p_2) \in \mathbb{R}^2$, the quantity $\frac{p_2 - p_1}{2}$ is the distance from $\mathbf{p}$ to the diagonal $x = y$. $\diamond$

Example 7.1.13 To understand what this means in practice, consider the persistence diagrams $X = \{\mathbf{p}_1, \mathbf{p}_2, \mathbf{p}_3, \mathbf{p}_4\}$ and $Y = \{\mathbf{q}_1, \mathbf{q}_2, \mathbf{q}_3, \mathbf{q}_4\}$ below

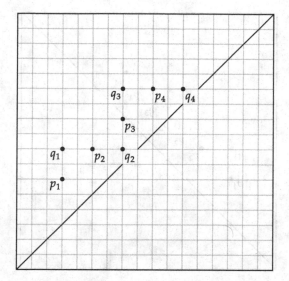

(a) For the empty partial match μ_0 where no points are matched, the cost is simply the max of the distances of the points (in both X and Y) to the diagonal, so $c(\mu_0) = 5$.

(b) For the partial match $\mu_1 = \mathbf{p}_i \mapsto \mathbf{q}_{5-i}$, the cost is maximized at the distance between $\mathbf{p}_1$ and $\mathbf{q}_4$, so $c(\mu_1) = 8$.

(c) For the partial match $\mu_2 = \mathbf{p}_i \mapsto \mathbf{q}_i$, the distances are all 2, so $c(\mu_2) = 2$.

It is easy to verify that in fact μ_2 minimizes the cost, hence $d_B(X, Y) = 2$.

This is all well and good, but we have not even verified that d_B is a metric. Before doing this, we ask "Where does the name "bottleneck" come from?" The next exercise will answer this.

Exercise 7.1.14 We consider sets of four points as in the previous example. The set X is unchanged, but in the set Y, the points $\{\mathbf{q}_2, \mathbf{q}_3, \mathbf{q}_4\}$ have changed position.

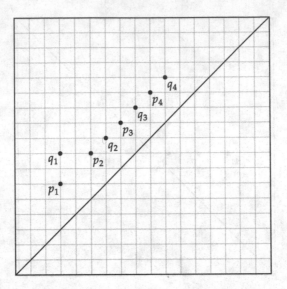

Compute $d_B(X, Y)$. Where does the name "bottleneck" come from?   ◇

Theorem 7.1.15 *The bottleneck distance d_B is a metric on persistence diagrams.*

Proof It follows from the definition that $d_B(X, X) = 0$ and

$$d_B(X, Y) = d_B(Y, X),$$

so the interesting part is the triangle inequality: we need to show that for three persistence diagrams $\{X, Y, Z\}$ that

$$d_B(X, Z) \leq d_B(X, Y) + d_B(Y, Z). \tag{7.1.4}$$

First, note that adding the diagonal to every persistence diagram means in particular that an unmatched point $\mathbf{p} = (p_1, p_2) \in X$ can be matched to a point of the diagonal of Y, and the distance is $\frac{p_2 - p_1}{2}$. This means that the three different subsets which need be considered in the computation of the cost of a partial match μ between X and Y as in Eq. 7.1.2 may be consolidated into the single condition

$$c(\mu) \;=\; \max\{\sup_{\mathbf{p} \in X}\{\|\mathbf{p} - \mu(\mathbf{p})\|_\infty\}\},$$

Now let μ_1, μ_2, and μ_3 be matchings that minimize the respective cost functions, so that

$$d_B(X, Z) \;=\; c(\mu_1)$$
$$d_B(X, Y) \;=\; c(\mu_2)$$
$$d_B(Y, Z) \;=\; c(\mu_3)$$

Now observe that we can compose partial matches: Suppose μ_{XY} is a partial match from $X \longrightarrow Y$ and μ_{YZ} is a partial match from $Y \longrightarrow Z$. Composing the two gives a partial match μ_{XZ}. Note that this can be the empty match, if the target Y' of μ_{XY} is disjoint from the source of μ_{YZ}. Hence

$$
\begin{aligned}
d_B(X, Z) \;&= c(\mu_1) \\
&= \max\{\sup_{\mathbf{p} \in X}\{||\mathbf{p} - \mu_1(\mathbf{p})||_\infty\}\} \\
&\leq \max\{\sup_{\mathbf{p} \in X}\{||\mathbf{p} - \mu_3\mu_2(\mathbf{p})||_\infty\}\} \\
&\leq \max\{\sup_{\mathbf{p} \in X}\{||\mathbf{p} - \mu_2(\mathbf{p})||_\infty\}\} = c(\mu_2) \\
&\quad + \max\{\sup_{\mathbf{q} \in Y}\{||\mathbf{q} - \mu_3(\mathbf{q})||_\infty\}\} = c(\mu_3) \\
&= d_B(X, Y) + d_B(Y, Z)
\end{aligned}
$$

In going from row 3 to row 4 in the argument, we are using the triangle inequality on the individual points

$$
||\mathbf{p} - \mu_3\mu_2(\mathbf{p})||_\infty \leq ||\mathbf{p} - \mu_2(\mathbf{p})||_\infty + ||\mu_2(\mathbf{p}) - \mu_3\mu_2(\mathbf{p})||_\infty
$$

What about long bars and the corresponding points on $\mathbb{R}_\infty^1$? Because we are dealing with a finite data set, by taking $\epsilon \gg 0$, all homology classes will have died, save a single component in H_0, which allows us to deal with this situation. □

7.2 Morse Theory

Persistent homology is similar to Morse theory, and to understand persistent homology it will be useful to have Morse theory as a backdrop. Morse developed the machinery almost a century ago in [115], [116]. We give a brief overview here in the spirit of [20] or [150]; for a more comprehensive treatment, see [108] or [111].

Definition 7.2.1 A smooth function $M \overset{f}{\to} \mathbb{R}$ on an n-dimensional compact manifold is defined on open sets $U \subseteq M$ having local coordinates $\{x_1, \ldots, x_n\}$. The function f is *Morse* if it has a finite number of critical points: those p where

$$
\frac{\partial f}{\partial x_1}(p) = \cdots = \frac{\partial f}{\partial x_n}(p) = 0,
$$

and the critical points are distinct and *nondegenerate*: the determinant of the *Hessian matrix*

$$
H_f = \begin{bmatrix} \frac{\partial^2 f}{\partial x_1^2}(p) & \cdots & \frac{\partial^2 f}{\partial x_1 \partial x_n}(p) \\ \vdots & \ddots & \vdots \\ \frac{\partial^2 f}{\partial x_n \partial x_1}(p) & \cdots & \frac{\partial^2 f}{\partial x_n^2}(p) \end{bmatrix}
$$

is nonzero at critical points. Morse theory generalizes the second derivative test.

Example 7.2.2 Flashback to vector calculus: the graph of $z = f(x, y)$ defines a surface S in $\mathbb{R}^3$. The critical points of f are the points p where

$$
\frac{\partial f}{\partial x}(p) = 0 = \frac{\partial f}{\partial y}(p),
$$

so the tangent plane $T_p(S)$ is horizontal and p is a max, min, or saddle point.

Exercise 7.2.3 Find the critical points and Hessians for the following functions

$$
z = x^2 + y^2, z = x^2 - y^2, z = -x^2 - y^2.
$$

Determine if the critical point is a max, min, or saddle.   ◇

A key ingredient in Morse Theory is the Morse Lemma:

Lemma 7.2.4 *For a Morse function f, if p is a critical point, then there exists a coordinate system $\{x_1, \ldots, x_n\}$ so that*

$$
f(x_1, \ldots, x_n) = f(p) - \sum_{i=1}^{\lambda_p} x_i^2 + \sum_{j=\lambda_p+1}^{n} x_j^2
$$

It follows immediately from the Morse Lemma that at a critical point p, the Hessian matrix is diagonal, with λ_p entries of -2, and $n - \lambda_p$ entries of $+2$. This means that λ_p is the number of negative eigenvalues of the Hessian at a critical point p.

Definition 7.2.5 The sublevel sets of a function f on M are $M_r = f^{-1}(-\infty, r]$.

Example 7.2.6 Let M be the torus, with $f : M \longrightarrow \mathbb{R}$ be the projection map defined by $f(x, y) = y$.

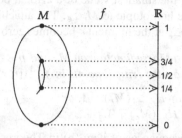

Note the topology of the sublevel sets M_r only changes at a finite number of points.

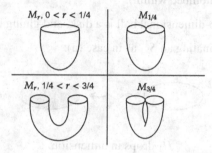

This agrees with our vector calculus intuition: if the tangent plane $T_p(S)$ to a smooth, nonplanar surface S is horizontal at p with equation $z = c$, then the plane $z = c + \epsilon$ either (locally near p) misses S, or intersects S in a curve. This dichotomy–that topological behavior of sublevel sets changes at critical points–is the content of the famous *Morse Theorem*:

Theorem 7.2.7 *Let $r < r' \in \mathbb{R}$ and f be a Morse function on M. Then*

(a) *If there are no critical points in $[r, r']$, then M_r and $M_{r'}$ are homotopic.*
(b) *If r' is the least value greater than r such that $p \in f^{-1}(r')$ is a critical point, then*

$$M_{r'} \simeq M_r \cup B_{\lambda_p},$$

with B_{λ_p} a ball of dimension λ_p, along with an attaching map sending

$$\partial(B_{\lambda_p}) \to M_r.$$

The homotopy in item (a) is a deformation retract: there is a continuous map $M_{r'} \to M_r$ leaving M_r fixed.

Example 7.2.8 A two-sphere punctured at the north pole deformation retracts to a point–the south pole–by sliding every point south along longitude lines.

If $r < r'$ with $r' = f(p)$ the smallest value of a critical point greater than r, then for $0 < \epsilon \le r' - r$, M_r is homotopic to $M_{r'-\epsilon}$. The attaching map can be thought of as gluing B_{λ_p} in to $M_{r'-\epsilon}$ at the moment ϵ becomes zero.

Theorem 7.2.9 *Let f be a Morse function on M, and $p \in f^{-1}(r')$ a critical point. Suppose $r < r'$ and there are no critical points in $f^{-1}([r, r'))$. Then either*

(a) $\dim H_{\lambda_p-1}(M_{r'}) = \dim H_{\lambda_p-1}(M_r) - 1$, *or*
(b) $\dim H_{\lambda_p}(M_{r'}) = \dim H_{\lambda_p}(M_r) + 1$.

Both possibilities are consequences of item (b) in Theorem 7.2.7. The intuition is that gluing in a ball B_{λ_p} can either fill in a $\lambda_p - 1$ dimensional hole, or be the final brick sealing off a hollow chamber of dimension λ_p (hopefully without Fortunato—*in pace requiescat*—entombed within).

Example 7.2.10 A two-dimensional ball is a disk, and gluing in a disk could

- Fill in a circle, eliminating an S^1, as in case (a):

H_1 drops in dimension.
- Put a lid on an open bowl, creating an S^2, as in case (b):

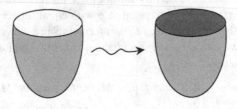

H_2 increases in dimension.

Corollary 7.2.11 *For a smooth manifold M and Morse function f, let c_i be the number of critical points p of f such that $\lambda_p = i$. Then*

$$c_i \ge \dim H_i(M).$$

Proof The number c_i bounds the number of i-dimensional cells in the CW complex used to construct M, hence also the ith homology.   □

Example 7.2.12 For the torus T^2 appearing in Example 7.2.6, at $r < \frac{1}{4}$, f is given locally by a function close to $x^2 + y^2$, and so $\lambda_p = 0$, similarly for $r > \frac{3}{4}$ the function f is given locally by a function close to $-x^2 - y^2$ so $\lambda_p = 2$. At the two saddle points H_f has one negative eigenvalue, so the vector of c_i is $(1, 2, 1)$. Compare this to the computation of the homology of T^2 in Example 5.3.4.

Definition 7.2.13 A continuous real valued function f on a space X has a homological critical value at q if the map induced by inclusion

$$\lim_{\epsilon \to 0} \left[H_p(f^{-1}(-\infty, q - \epsilon)) \to H_p(f^{-1}(-\infty, q + \epsilon)) \right]$$

is not an isomorphism. The function f is *tame* if it has a finite number of homological critical values, and the homology $H_p(f^{-1}(-\infty, q], \mathbb{K})$ of the sublevel sets is finite dimensional for all points $q \in X$.

For example, if X is smooth and f is a Morse function then the homological critical points are the usual critical points of f. Tame functions are the right setting for the main result of the next section.

7.3   The Stability Theorem

For a manifold M and smooth real valued function $f : M \to \mathbb{R}$, the sublevel sets $M_r = f^{-1}(-\infty, r]$ appearing in Sect. 7.2 yield a filtration of M. In this section, we outline work of Cohen-Steiner–Edelsbrunner–Harer [44], which proves a *stability* theorem for persistent homology; [57] calls their result "arguably the most powerful theorem in TDA". The motivation is natural: for persistent homology to be useful, if two spaces X and Y are close, then their corresponding persistence diagrams should also be close. Recall that

Definition 7.3.1 For functions $f, g : X \to \mathbb{R}$, the sup or L_∞ norm is

$$\|f - g\|_\infty = \max_{x \in X} |f(x) - g(x)|.$$

Let $f : X \to \mathbb{R}$ and let $D(f)$ be the persistence diagram of some homology module of the filtration on X induced by f. The main theorem of [44] is

Theorem 7.3.2 (Stability Theorem) *Let X be a triangulable space and f, g continuous tame real valued functions on X. Then the persistence diagrams of f and g satisfy $d_B(D(f), D(g)) \le \|f - g\|_\infty$.*

To tackle stability, we proceed as follows: first, suppose that X is a manifold and f, g are Morse functions. By Theorem 7.2.9 of the previous section, there are a finite number of critical points, and in this context the main result on stability is the statement that the bottleneck distance between the persistence diagrams of f and g is bounded by the L_∞ distance between the functions f and g themselves.

The proof by Cohen-Steiner–Edelsbrunner–Harer in [44] involves a fairly intricate diagram chase. We instead give the proof due to Cohen-Steiner–Edelsbrunner–Morozov [46] of a slightly weaker version: Theorem 7.3.7 below applies in the setting of simplicial complexes and monotone functions, so it is well suited to handle the Čech or Rips complex arising from point cloud data. Before moving to the formal proof, we sketch an intuitive example

Example 7.3.3 Consider the graph of $y = f(x)$ below. We take as our filtered topological space $X_y = \{x \in \mathbb{R}^1 \mid f(x) \leq y\}$.

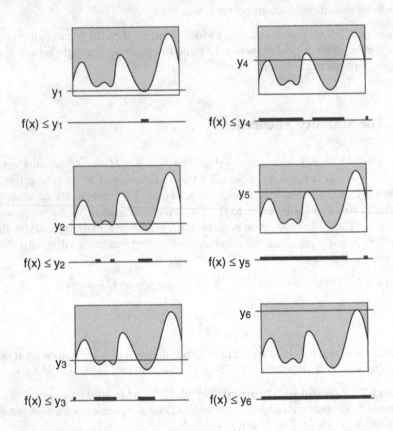

Let $\{y_1 < y_2 < \cdots < y_6\}$ be the y-values where a relative min or max occurs. At y_3, two components of X_y merge, and a new component appears (on the extreme left). It follows from Definition 7.1.2 that when two classes merge, the class which is born later has image zero under the induced map–so in the persistence diagram, the class born earlier "survives" (this is known as the *elder rule*). In the merge occurring at y_3, the classes were both born at y_2, so there is no distinction to worry about. Hence the persistence diagram for $H_0(X_y)$ is

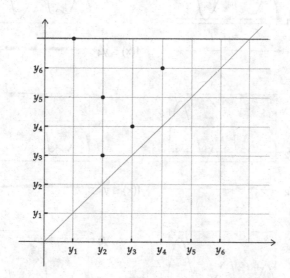

Example 7.3.4 When there is a small perturbation in the value of a function, this can cause a class to appear in $H_0(X_y)$, as below. Note that in this example, the new class has a very short lifespan.

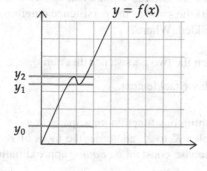

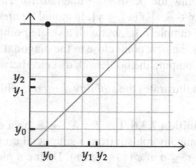

Exercise 7.3.5 Suppose the function appearing in Example 7.3.3 has a small perturbation, as in the figure below. Note that there are now two new points where a relative min or max occurs; call them y_3' and y_3'', with $y_3 < y_3' < y_3'' < y_4$.

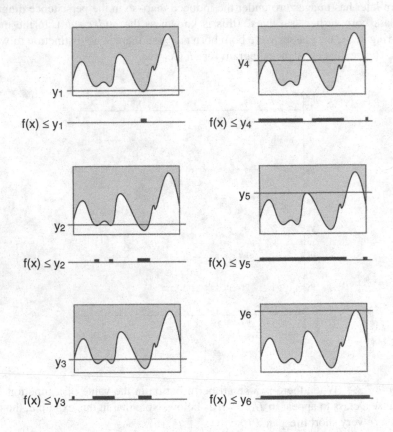

- Draw the persistence diagram for H_0 of the filtered topological spaces $X_y = \{x \in \mathbb{R}^1 \mid f(x) \leq y\}$. It should be *almost* the same as the persistence diagram of Example 7.3.3, but with a single point added. Where?
- Is the new point close to the diagonal?
- Compute the bottleneck distance between the two persistence diagrams.

This illustrates the *stability theorem*, which we tackle next. ◇

Definition 7.3.6 Let $f : X \to \mathbb{R}$ be a continuous function on a space, $K \xrightarrow{\phi} X$ a homeomorphism from a simplicial complex K to X, and order the faces of K so that $\tau \leq \sigma$ if $\phi(\tau) \subseteq \phi(\sigma)$. Define a piecewise constant *monotone* approximation $\overline{f}$ to f such that $\overline{f}(\tau) \leq \overline{f}(\sigma)$ if $\tau \leq \sigma$.

For example, one way to do this is via $\sigma \mapsto \max\{f(x) \mid x \in \sigma\}$. An important ingredient in the proof below is the Morse-like pairing between simplices of [65] described in Sect. 7.1.3. Define a k−simplex τ to be *positive* if it belongs to a k−cycle, and *negative* otherwise. The kth betti number is then the number of positive k-classes (births), minus the number of negative $(k + 1)$−classes (deaths). Hence a birth-death pair is encoded in a pair (τ, σ) with $\tau \subseteq \sigma$, and τ positive and σ negative.

Theorem 7.3.7 (Combinatorial Stability Theorem, [46]) *For monotone $\overline{f}, \overline{g}$ mapping a simplical complex $K \rightarrow \mathbb{R}$,*

$$d_B(D(\overline{f}), D(\overline{g})) \leq ||\overline{f} - \overline{g}||_\infty.$$

Proof Let σ be a face of K. Since $\overline{f}$ and $\overline{g}$ are monotone, the homotopy

$$h_t(\sigma) = (1 - t)\overline{f}(\sigma) + t\overline{g}(\sigma) \text{ for } t \in [0, 1]$$

is also monotone; since h_t is monotone it induces an ordering on the simplices of K. Say the ordering changes at $\{t_1, \ldots, t_k\}$, and put $t_0 = 0, t_{k+1} = 1$. For $i \in \{0, \ldots, k\}$, take r and s such that $t_i \leq r < s < t_{i+1}$. If (τ, σ) is a birth-death pair of simplices defined for the ordering that exists for $r \leq t \leq s$, then

$$(h_r(\tau), h_r(\sigma)) \in D(h_r) \text{ and } (h_s(\tau), h_s(\sigma)) \in D(h_s).$$

From the definition of the L_∞ distance, we then have

$$
\begin{aligned}
d_B(D(h_r), D(h_s)) &\leq & ||h_r - h_s||_\infty \\
&=& (s - r) \cdot ||\overline{f} - \overline{g}||_\infty.
\end{aligned}
$$

The authors of [46] show that transposing a pair of simplices, while it changes the pairing, does not change the persistence diagram. Therefore,

$$
\begin{aligned}
d_B(D(\overline{f}), D(\overline{g})) &\leq & \sum_{i=0}^{k} d_B(D(h_{t_i}), D(h_{t_{i+1}})) \\
&\leq & \left(\sum_{i=0}^{k} (t_{i+1} - t_i) \right) ||\overline{f} - \overline{g}||_\infty.
\end{aligned}
$$

Since $\sum_{i=0}^{k} (t_{i+1} - t_i) = 1$, this concludes the proof.   □

As noted, the proof above is for a somewhat special situation. The notion of interleaving appearing in Definition 7.4.7 plays a key role in many versions of stability. We discuss this in the next section, and close with an application of [114] applying interleaving to merge trees.

7.4 Interleaving and Categories

Persistent Homology fits naturally into a category theoretic framework, and this viewpoint will be useful in formulating the concept of *interleaving distance*. The reason to bring interleaving distance into the picture is that it provides a natural way to define a metric on spaces related to diagrams.

7.4.1 Categories and Functors

Category Theory is an abstract framework that encompasses homology, persistent homology, and constructions such as Hom, $\otimes$, and localization. Used judiciously, category theory provides a powerful tool to streamline and unify many mathematical constructions, see MacLane [107] or Riehl [128] for detailed expositions.

In this section we'll see that persistent homology can be regarded as a *functor* from the poset category of the real numbers $(\mathbb{R}, \leq)$ to the category of vector spaces over a chosen field $\mathbb{K}$. We begin by giving a brief review of categories and functors; the material is developed in depth when we introduce derived functors in Chap. 9. Then we define the interleaving distance, and show that it does define a metric. For a detailed exposition on interleaving and stability, see the monograph [40] by Chazal–de Silva–Glisse–Oudot.

Definition 7.4.1 A *Category* $\mathscr{A}$ consists of

- A collection of objects $\mathrm{Ob}(\mathscr{A})$.
- A collection of morphisms: for $A, B \in \mathrm{Ob}(\mathscr{A})$, a set $\mathrm{Hom}(A, B)$,

such that morphisms can be composed. The composition is associative, and for each object A, there is an identity morphism $1_A : A \to A$.

Example 7.4.2 Examples of categories

- **Set**: Sets and functions.
- **P**: Posets and order preserving functions.
- **Vect**: Vector spaces and linear transformations.
- **Top**: Topological spaces and continuous maps.

In many mathematical settings we deal with objects and maps between them. A map between categories is

Definition 7.4.3 A *functor* F is a function from a category $\mathscr{A}$ to a category $\mathscr{B}$, which takes objects to objects $\mathrm{Ob}(\mathscr{A}) \overset{F}{\Rightarrow} \mathrm{Ob}(\mathscr{B})$ and morphisms to morphisms in a way that preserves identities and compositions. The functor F is *covariant* if

$$F : A \overset{f}{\to} A' \in \mathrm{Hom}(A, A') \implies F(A) \overset{F(f)}{\to} F(A') \in \mathrm{Hom}(F(A), F(A')),$$

and *contravariant* if

$$F : A \overset{f}{\to} A' \in \mathrm{Hom}(A, A') \implies F(A') \overset{F(f)}{\to} F(A) \in \mathrm{Hom}(F(A'), F(A)).$$

Exercise 7.4.4 For a commutative ring R, show the set of R-modules and R-module homomorphisms is a category R-mod, and for a fixed module N, that

(a) $\mathrm{Hom}_R(N, \bullet)$ is covariant, and $\mathrm{Hom}_R(\bullet, N)$ is contravariant.
(b) $\bullet \otimes_R (N)$ is covariant.

What happens when composing functors?   ◇

The functors above take R-mod to R-mod. Functors are most interesting when they provide a way to translate between categories that are of very different flavors.

Example 7.4.5 Homology is a functor for any of the homology theories of Chaps. 4 and 5; for example, singular homology takes **Top** to **Vect**:

- if $X \xrightarrow{1_X} X$ then $H_i(X) \xrightarrow{H(1_X)} H_i(X)$.
- if $X \xrightarrow{f} Y$ and if $Y \xrightarrow{g} Z$ then $H_i(g \circ f) = H_i(g) \circ H_i(f)$.

Exercise 7.4.6 Connect the two concepts below using category theory.

(a) In Definition 3.1.9, we defined a homotopy of maps: two continuous maps of topological spaces $f_0, f_1 : S \to T$ are *homotopic* if there is a continuous map F such that

$$S \times [0, 1] \xrightarrow{F} T$$

with $F(x, 0) = f_0(x)$ and $F(x, 1) = f_1(x)$; F deforms f_0 to the map f_1.
(b) In Definition 4.3.7, we defined a chain homotopy of complexes: given chain maps

$$A_i \xrightarrow{\alpha_i} B_i \text{ and } A_i \xrightarrow{\beta_i} B_i$$

a chain homotopy is a map

$$A_i \xrightarrow{\gamma_i} B_{i+1}$$

such that

$$\alpha_i - \beta_i = \gamma_{i-1}\delta_i + \partial_{i+1}\gamma_i$$

in the diagram below

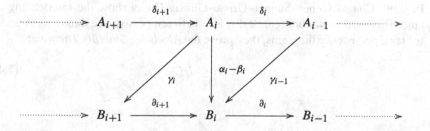

Theorem 4.3.8 shows that chain homotopic maps induce the same map on homology. Prove that homotopic maps of type (a) induce homotopic maps of type (b), and show Theorem 4.2.11 follows as a corollary.   ◇

7.4.2 Interleaving

In Exercise 7.4.6 we revisited homotopies between chain complexes and connected them to homotopies of continuous maps. *Interleaving* is a variation on this theme, introduced by Chazal–Cohen-Steiner–Glisse–Guibas–Oudot in [39]. Roughly speaking, interleaving is a way of sandwiching sequences between each other. As in the "squeeze" theorem on limits encountered in freshman calculus, this means the sequences under consideration are close.

Definition 7.4.7 Let $\mathscr{C}$ be a category and $A_\bullet$, $B_\bullet$ be sets of objects of $\mathscr{C}$, with maps indexed by pairs of related elements of a poset P. For us, the poset P will be $(\mathbb{R}, \leq)$ or $(\mathbb{Z}, \leq)$, with A and B families of persistent homology modules as in Definition 7.1.2. An ϵ-*interleaving* is a sequence of morphisms

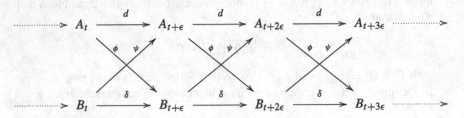

such that the diagrams commute:

$$d^2 = \psi \circ \phi \text{ and } \delta^2 = \phi \circ \psi.$$

Define $d_I(A, B) = \inf\{\epsilon \in P \cup \infty \mid A_\bullet \text{ and } B_\bullet \text{ are } \epsilon\text{-interleaved}\}$.

Exercise 7.4.8 Show that d_I defines an extended pseudometric, the *interleaving distance*. A pseudometric satisfies the same properties of a metric, with a caveat: the distance between two distinct points can be zero. The adjective *extended* comes from including ∞; the triangle inequality only applies if the distances are finite. ◇

In [39], Chazal–Cohen-Steiner–Glisse–Guibas–Oudot show the interleaving distance is related to the bottleneck distance between persistence diagrams A and B: for tame persistence diagrams, they prove the *Algebraic Stability Theorem*:

$$d_B(A, B) \leq d_I(A, B). \tag{7.4.1}$$

The Isometry Theorem. Lesnick shows in [103] that the converse of Eq. 7.4.1 holds; the resulting equality is known as the *Isometry Theorem*. We sketch a proof of the isometry theorem following the approach of Bauer-Lesnick in [7]; this also provides an opportunity to introduce the concept of a *persistence module*, a central object of topological data analysis.

Definition 7.4.9 A *persistence module* X is a functor from a poset P to the category $\mathbf{Vect}_{\mathbb{K}}$; X is called *pointwise finite dimensional* (p.f.d) if it takes values in finite dimensional vector spaces. So if $x_1 \preceq x_2 \preceq x_3$ in P, then there are linear transformations of the vector spaces $X(x_i)$

$$X(x_1) \xrightarrow{X_{\leq 21}} X(x_2) \xrightarrow{X_{\leq 32}} X(x_3), \tag{7.4.2}$$

such that $X_{\leq ii}$ is the identity, and $X_{\leq 32} \circ X_{\leq 21} = X_{\leq 31}$.

The conditions of Eq. 7.4.2 echo the compatibility conditions we've seen defined in a number of different settings, ranging from direct limits to transition functions for vector bundles.

The role of X in Definition 7.4.9 will be played by the homology of a filtered simplicial complex. We need a bit more terminology: a morphism of persistence modules X and Y is given by a family of maps such that all diagrams of the form below commute

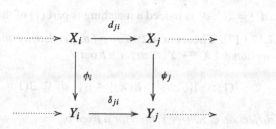

Notation The ϵ-*shift* $X(\epsilon)$ of a persistence module X is defined exactly as is the shift of graded modules in Definition 6.4.6: $X(\epsilon)_i = X_{\epsilon+i}$; we say X is ϵ-*trivial* if the map $X \to X(\epsilon)$ is zero. For I a connected subset of $\mathbb{R}$, define the *interval persistence* module $C(I)_t = \mathbb{K}$ if $t \in I$, and 0 otherwise.

We call X *interval decomposable* if $X \simeq \oplus_{I \in S} C(I)$ for a finite set S; in this case the finite set S is the barcode of X. We write $\langle b, d \rangle$ for an interval $b < d$ (open, closed, or neither—see Sect. 2.3 of [7] for more precision). We can write the barcode of an interval decomposable module as a finite disjoint union of intervals with fixed right endpoints, or a disjoint union of intervals with fixed left endpoints:

$$B(X) = \bigsqcup_{d \in P \cup \infty} \langle \cdot, d \rangle = \bigsqcup_{b \in P \cup -\infty} \langle b, \cdot \rangle. \tag{7.4.3}$$

Example 7.4.10 For the interval modules $X = C(\langle b, d \rangle)$ and $Y = C(\langle b', d' \rangle)$, suppose $d' - d = \epsilon$. If ϵ is positive, there is no morphism from X to Y, because shifting d right in X by $\frac{\epsilon}{2}$ yields zero, whereas pushing d forward to Y and shifting right by $\frac{\epsilon}{2}$ does not yield zero in Y. There can be a morphism if we allow shifting.

Exercise 7.4.11 Show if $X = C([b, d))$ and $Y = C([b', d'))$ then there is a morphism $X \to Y$ iff $b' \leq b < d' \leq d$: the endpoints of the bar representing Y lie (weakly) to the left of the respective endpoints of the bar representing X. Note the swap from $\langle b, d \rangle$ to $[b, d)$–there is no morphism from $[0, 1)$ to $[0, 1]$.   ◇

The proof of the Isometry Theorem in [7] begins with a structure result on submodules and quotient modules of persistence modules: Theorem 7.4.12 below. Then for a morphism f of persistence modules, define a matching $\mu_f : B(X) \to B(Y)$ as follows. First, suppose f is an inclusion, and the set of bars in Eq. 7.4.3 with right endpoint d is

$$\langle b_1, d \rangle \supseteq \langle b_2, d \rangle \supseteq \langle b_3, d \rangle \cdots$$

and the set of bars of $B(Y)$ with right endpoint d is

$$\langle b'_1, d \rangle \supseteq \langle b'_2, d \rangle \supseteq \langle b'_3, d \rangle \cdots$$

That $\mu_f(\langle b_i, d \rangle) = \langle b'_i, d \rangle$ is indeed a matching is part (1) of the next theorem.

Theorem 7.4.12 ([7], Theorem 4.2) *For p.f.d. persistence modules X and Y, if there is an injection $\phi : X \hookrightarrow Y$, then for a fixed d,*

$$(1): \quad |\{\langle \cdot, d \rangle \subseteq B(X)| \leq |\{\langle \cdot, d \rangle \subseteq B(Y)|.$$

If there is a surjection $\phi : X \twoheadrightarrow Y$ then for a fixed b,

$$(2): \quad |\{\langle b, \cdot \rangle \subseteq B(X)| \geq |\{\langle b, \cdot \rangle \subseteq B(Y)|.$$

(1) yields a map $B(X) \hookrightarrow B(Y)$ via $\langle b_i, d \rangle \mapsto \langle b'_i, d \rangle$ for $i \leq |\{\langle \cdot, d \rangle \subseteq B(X)|$.
(2) yields a map $B(Y) \hookrightarrow B(X)$ via $\langle b, d'_i \rangle \mapsto \langle b, d_i \rangle$ for $i \leq |\{\langle b, \cdot \rangle \subseteq B(Y)|$.

Definition 7.4.13 A matching of barcodes μ is called an ϵ-matching if

(a) $B(X)_{2\epsilon} \subseteq \mathrm{coim}(\mu)$ and $B(Y)_{2\epsilon} \subseteq \mathrm{im}(\mu)$.
(b) If $\mu(\langle b, d \rangle) = \langle b', d' \rangle$, then $\langle b, d \rangle \subseteq \langle b' - \epsilon, d' + \epsilon \rangle$ and $\langle b', d' \rangle \subseteq \langle b - \epsilon, d + \epsilon \rangle$.

So $d_B(B(X), B(Y)) = \inf\{\epsilon \mid \exists \text{ an } \epsilon - \text{matching between } B(X) \text{ and } B(Y)\}$. Our aim is to produce an ϵ-matching from an ϵ-interleaving. Let $B(Z)_\gamma$ denote the bars of $B(Z)$ of length $\geq \gamma$, and let $\mathrm{coim}(\mu), \mathrm{im}(\mu)$ be as in Definition 7.1.10.

Theorem 7.4.14 (Induced Matching, [7], Theorem 6.1) *For a map of p.f.d. modules $f : X \to Y$, suppose $\mu_f(\langle b, d \rangle) = \langle b', d' \rangle$, then*

- *If* $\mathrm{coker}(f)$ *is ϵ-trivial, then all intervals of length $\geq \epsilon$ in $B(Y)$ are contained in* $\mathrm{im}(\mu_f)$ *and* $b' \leq b \leq b' + \epsilon$.
- *If* $\ker(f)$ *is ϵ-trivial, then all intervals of length $\geq \epsilon$ in $B(X)$ are contained in* $\mathrm{coim}(\mu_f)$ *and* $d - \epsilon \leq d' \leq d$.

An interleaving of persistence modules $f : X \longrightarrow Y$ (indeed, any map) factors as

$$X \twoheadrightarrow \mathrm{im}(f) \hookrightarrow Y. \qquad (7.4.4)$$

The proof of the Isometry Theorem hinges on using this factorization to match $B(\mathrm{im}(f))$ with both $B(X)$ and $B(Y)$. The next example (5.2 of [7]) illustrates that the induced matching obtained from the factorization is somewhat subtle.

Example 7.4.15 Let $X = C(\langle 1, 2 \rangle) \oplus C(\langle 1, 3 \rangle)$ and $Y = C(\langle 0, 2 \rangle) \oplus C(\langle 3, 4 \rangle)$, with $f : X \to Y$ injecting $C(\langle 1, 2 \rangle) \hookrightarrow C(\langle 0, 2 \rangle)$ and $C(\langle 1, 3 \rangle) \mapsto 0$. Therefore the map f factors as

$$C(\langle 1, 2 \rangle) \oplus C(\langle 1, 3 \rangle) \longrightarrow \mathrm{im}(f) = C(\langle 1, 2 \rangle) \longrightarrow C(\langle 0, 2 \rangle) \oplus C(\langle 3, 4 \rangle).$$

Algorithm: (1) match right endpoints (2) find the image (3) match left endpoints. Using (2) of Theorem 7.4.12, when matching left endpoints, we match longer intervals to longer intervals. So matching $C(\langle 1, 2 \rangle)$ to $C(\langle 0, 2 \rangle)$ is wrong; μ sends $C(\langle 1, 3 \rangle)$ to $C(\langle 0, 2 \rangle)$. Notice that matching does not respect direct sum.

Theorem 7.4.16 *For X, Y p.f.d. modules, $d_B(B(X), B(Y)) = d_I(X, Y)$.*

Sketch that $d_B \leq d_I$: Let r_ϵ denote a right-shift of barcodes by ϵ, so that the map $r_\epsilon : B(Y(\epsilon)) \to B(Y)$ is a bijection. Note that if $\epsilon > 0$ there is no nonzero map $Y(\epsilon) \to Y$, so r_ϵ does not have a functorial interpretation. If $f : X \to Y$ is an ϵ-interleaving, with $\mu_f(\langle b, d \rangle) = \langle b', d' \rangle$, then it follows from the definition that $\ker(f)$ and $\mathrm{coker}(f)$ are 2ϵ-trivial.

Applying Theorem 7.4.14, we have that all intervals of length $\geq 2\epsilon$ in $B(Y)$ are contained in $\mathrm{im}(\mu_f)$ and that $b' \leq b \leq b' + 2\epsilon$, and that all intervals of length $\geq 2\epsilon$ in $B(X)$ are contained in $\mathrm{coim}(\mu_f)$ and that $d - 2\epsilon \leq d \leq d'$. Composing this with the shift operator, we see that $r_\epsilon \circ \mu_f$ is an ϵ-matching.

Sketch that $d_I \leq d_B$: Start by fixing ϵ, and suppose μ is an ϵ-matching; we need to produce an ϵ-interleaving. Let $X_\epsilon = \oplus_{I \in \mathrm{coim}(\mu)} C(I)$ and $Y_\epsilon = \oplus_{I \in \mathrm{im}(\mu)} C(I)$, and write persistence modules X and Y as

$$X = X_\epsilon \oplus (X \setminus X_\epsilon) \text{ and } Y = Y_\epsilon \oplus (Y \setminus Y_\epsilon),$$

By condition (b) in Definition 7.4.13, for a matched pair I, I' we have

$$I \xrightarrow{\mu} I' \text{ with } I \subseteq N_\epsilon(I') \text{ and } I' \subseteq N_\epsilon(I).$$

Hence we can define an interleaving between these sets, which takes care of intervals longer than ϵ. Since $X \setminus X_\epsilon$ and $Y \setminus Y_\epsilon$ have summands which are interval modules with bars shorter than ϵ, they are in (respectively) the kernel and cokernel of μ, and we define an interleaving by sending these intervals to zero. $\square$

Corollary 7.4.17 ([7]) *Two p.f.d. persistence modules X and Y are ϵ-interleaved iff there exists a morphism $f : X \to Y(\epsilon)$ such that all intervals in $B(\ker f)$ and $B(\operatorname{im} f)$ are of length at most 2ϵ.*

History and Further Developments The original proof of [39] is fairly intricate; there are three main steps. First, discretize the problem. Second, prove a relation between the points contained in certain boxed regions of the persistence diagrams A and B (the "Box Lemma"). Conclude by setting up a sequence of objects which interpolate between A and B.

Chazal–de Silva–Glisse–Oudot give a detailed treatment of stability and related topics in [40], touching on Bubenik's description of the interpolating step as a Kan extension. Work on fitting interleaving into a broader categorical framework appears in papers of Bubenik-de Silva–Scott in [22], [23].

An important aspect of interleaving is that it provides (see [103]) a natural metric in the *multiparameter* setting. See de Silva–Munch–Stefanou [57] for work on interleaving in categories with a flow, Bjerkevik–Botnan–Kerber [13], [14] and Chachólski–Gäfvert [35] for results on computational complexity of interleaving, and Dey–Fan–Wang [58] for a version where the horizontal maps are not inclusions. A detailed exposition of Bauer-Lesnick's proof also appears in [126], which applies persistence to study questions in symplectic geometry.

7.4.3   Interleaving Vignette: Merge Trees

An illustrative use of interleaving distance is the work of Morozov–Beketayev–Weber [114] on *Merge Trees*. A merge tree is an encoding of the connected components of the sublevel sets of a function–an edge traces the evolution of a component over time, with two edges meeting in a vertex when the connected components merge.

Definition 7.4.18 For a function $X \xrightarrow{f} \mathbb{R}$,

- the *Reeb graph* is a quotient space $X/\sim$, formed as in Definition 3.1.14 of Chap. 3: let $x_1 \sim x_2$ if they lie in the same connected component of

$$f^{-1}(f(x_1)) = f^{-1}(f(x_2))$$

So for a surface, quotienting contracts every contour line to a point.

- the *merge tree* T_f is a rooted graph which tracks the connected components of the sublevel sets of f. Formally, let

$$\mathrm{Up}(f) = \{(x, y) \in X \times \mathbb{R} \mid y \geq f(x)\}.$$

Project $\mathrm{Up}(f) \to \mathbb{R}$ via $\overline{f}(x, y) = y$; the Reeb graph of $\overline{f}$ is the merge tree T_f. There is a well-defined ancillary map $\widehat{f} : T_f \to \mathbb{R}$ sending $x \in \mathrm{Up}(f) \mapsto y$, where y is the component of $\overline{f}^{-1}(\overline{f}(x))$ containing x. Note that the projection π of a level set of $\overline{f}$ in $\mathrm{Up}(f)$ to X is a sublevel set of the original function f. Let $F_p = \pi(\overline{f}^{-1}(p)) = f^{-1}(-\infty, p]$.

- for $\epsilon \in \mathbb{R}$, the inclusion $F_p \hookrightarrow F_{p+\epsilon}$ means that a connected component X of F_p is contained in a connected component $Y \subseteq F_{p+\epsilon}$. This allows us to define the ϵ-*shift map* on the merge tree

$$T_f \xrightarrow{i^\epsilon} T_f$$

as follows: $x \in T_f$ corresponds to a connected component X as above, which maps to Y, which corresponds to y in T_f; define $i^\epsilon(x) = y$.

The intuition is that i^ϵ moves $x \in T_f$ towards the root of T_f, and stops when it reaches y corresponding to $Y \subseteq F_{p+\epsilon}$ as above.

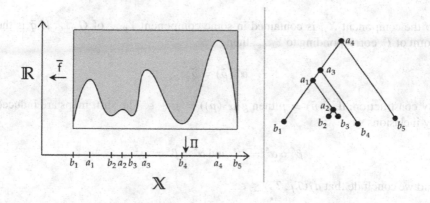

In order to relate a pair of merge trees T_f and T_g, we consider interleaving maps between the two. Continuous maps

$$T_f \xrightarrow{\alpha^\epsilon} T_g \text{ and } T_g \xrightarrow{\beta^\epsilon} T_f$$

are ϵ-compatible if they satisfy the criteria of Definition 7.4.7, where

$$\widehat{g}(\alpha^\epsilon(x)) = \widehat{f}(x) + \epsilon \quad \widehat{f}(\beta^\epsilon(y)) = \widehat{g}(y) + \epsilon$$
$$\beta^\epsilon \circ \alpha^\epsilon = i^{2\epsilon} \qquad \alpha^\epsilon \circ \beta^\epsilon = j^{2\epsilon},$$

with $T_f \xrightarrow{i^{2\epsilon}} T_f$ and $T_g \xrightarrow{j^{2\epsilon}} T_g$ are the 2ϵ-shift maps in their trees. The interleaving distance on merge trees comes from Definition 7.4.7:

$$d_I(T_f, T_g) = \inf\{\epsilon \mid \text{there exist } \epsilon\text{-compatible maps as above.}\}$$

Here are the two main results of [114] on interleaving distance for merge trees:

Theorem 7.4.19 *For $f, g : X \longrightarrow \mathbb{R}$, the merge trees T_f and T_g satisfy*

$$d_I(T_f, T_g) \le \|f - g\|_\infty.$$

Proof We interleave the sublevel sets of f and g. If $\|f - g\|_\infty = \epsilon$ then since $F_p = f^{-1}(-\infty, p]$ and $G_q = g^{-1}(-\infty, q]$,

$$F_p \subseteq G_{p+\epsilon} \subseteq F_{p+2\epsilon}.$$

Now,

$$\tilde{p} \in T_f \text{ corresponds to a component } X_p \subseteq F_p \subseteq G_{p+\epsilon},$$

so the component X_p is contained in some component $Y_{p+\epsilon}$ of $G_{p+\epsilon}$. If $\tilde{q}$ is the point of T_g corresponding to $Y_{p+\epsilon}$, then

$$\alpha^\epsilon(\tilde{p}) = \tilde{q}.$$

By construction, if $\widehat{f}(\tilde{p}) = p$ then $\widehat{g}(\alpha^\epsilon(\tilde{p})) = p + \epsilon$. The shift maps are induced by inclusions so

$$\beta^\epsilon \circ \alpha^\epsilon = i^{2\epsilon} \text{ and } \alpha^\epsilon \circ \beta^\epsilon = j^{2\epsilon}$$

and we conclude that $d_I(T_f, T_g) \le \epsilon$. $\qquad\qquad\qquad\qquad\qquad\qquad\qquad\qquad$ $\square$

There is a similar result for bottleneck distance. The proof uses a version of the result of [39], simplified to the setting $p = 0$. Let $f, g : X \longrightarrow \mathbb{R}$ be tame functions such that there are ϵ-interleavings as below

$$H_0(F_t) \xrightarrow{\phi^\epsilon} H_0(G_{t+\epsilon}) \text{ and } H_0(G_t) \xrightarrow{\psi^\epsilon} H_0(F_{t+\epsilon}),$$

which commute with the inclusion maps. Then

$$d_B(D_0(f), D_0(g)) \le \epsilon.$$

Theorem 7.4.20 *For $f, g : X \longrightarrow \mathbb{R}$, the merge trees T_f and T_g satisfy*

$$d_B(D_0(f), D_0(g)) \leq d_I(T_f, T_g)$$

Proof Because merge trees contract sublevel sets to points and H_0 measures the number of connected components,

$$X \xrightarrow{f} \mathbb{R} \text{ and } T_f \xrightarrow{\widehat{f}} \mathbb{R}$$

have the same zero-dimensional persistence diagrams

$$D_0(f) = D_0(\widehat{f}).$$

Hence it suffices to show that

$$d_B(D_0(\widehat{f}), D_0(\widehat{g})) \leq d_I(T_f, T_g) = \epsilon.$$

Since $d_I(T_f, T_g) = \epsilon$, we have maps

$$T_f \xrightarrow{\alpha^\epsilon} T_g \text{ and } T_g \xrightarrow{\beta^\epsilon} T_f$$

which commute with the induced inclusion maps. So in particular, $\widehat{f}^{-1}(-\infty, a]$ and $\widehat{g}^{-1}(-\infty, a]$ are ϵ-interleaved, hence so are the corresponding groups

$$H_0(\widehat{f}^{-1}(-\infty, a]) \text{ and } H_0(\widehat{g}^{-1}(-\infty, a]).$$

Applying the variant of [39] described above concludes the proof. □

7.4.4 Zigzag Persistence and Quivers

We close with another connection between TDA and category theory. Our standard setup thus far has involved fattening the dots in a point cloud, yielding a filtered topological space, with

$$X_\epsilon \hookrightarrow X_{\epsilon'} \text{ when } \epsilon \leq \epsilon'.$$

What about the case where we have a parameterized family of topological spaces that can both dilate and contract? Such situations arise naturally: suppose for example the parameter ϵ represents time, and X_ϵ is the aorta. To deal with such situations, we are led to

$$X_\epsilon \longleftrightarrow X_{\epsilon'} \text{ when } \epsilon \leq \epsilon',$$

where $\longleftrightarrow$ means either $X_\epsilon \leftarrow X_{\epsilon'}$ or $X_\epsilon \rightarrow X_{\epsilon'}$. Passing to the level of vector spaces, we have a diagram

$$V_1 \longleftrightarrow V_2 \longleftrightarrow V_3 \longleftrightarrow \cdots V_n \qquad (7.4.5)$$

In [26] Carlsson–de Silva introduce *Zigzag Persistence* to handle cases of this type: it turns out that the theory of quiver representations provides the perfect framework for addressing the situation. We close by giving a brief sketch of the ingredients that go into developing the theory of zigzag persistence.

Definition 7.4.21 A *quiver* is a directed graph as in Definition 3.4.1 (note that loops and multiple edges are allowed). A *quiver representation* is an assignment of a vector space to each vertex, and a linear map to each directed edge.

Given two quiver representations V and W, the direct sum representation $V \oplus W$ is formed in the obvious way. A quiver representation is *decomposable* if it is isomorphic to a direct sum of nonzero quiver representations. As in group representation theory, a first question about quivers is "What are the fundamental building blocks?", that is, what are the indecomposable representations of a quiver? A quiver is said to be of *finite type* if there only finitely many (up to isomorphism) classes of indecomposable representations.

Theorem 7.4.22 (Gabriel,   [77]) *A quiver is of finite type iff the associated undirected graph corresponds to a Dynkin diagram of type ADE.*

The classification of complex simple Lie algebras is a gem of mathematics: there are four infinite families and five sporadic cases; each can be compactly encoded by a *Dynkin diagram*, which is an undirected graph. The diagrams of type ADE are simple graphs, with n denoting the number of vertices.

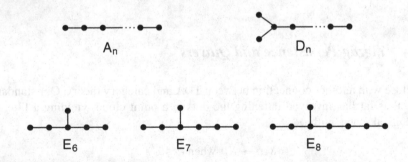

By Gabriel's theorem, a quiver of type A has finite representation type, which means that the diagram of vector spaces in Eq. 7.4.5 can be decomposed, for any given instantiation of the individual $\longleftrightarrow$ symbols into $\longleftarrow$ or $\longrightarrow$, into a finite set of irreducibles. Carlsson–de Silva first show that in the TDA setting the indecomposables correspond to intervals with fixed birth and death, and then go on to develop tools to relate zigzag diagrams such as a sequence of unions

$$X_1 \longrightarrow X_1 \cup X_2 \longleftarrow X_2 \longrightarrow X_2 \cup X_3 \longleftarrow X_3 \cdots$$

and of intersections

$$X_1 \longleftarrow X_1 \cap X_2 \longrightarrow X_2 \longleftarrow X_2 \cap X_3 \longrightarrow X_3 \cdots$$

Zigzag persistence has proved to be an important and flexible addition to the TDA toolbox; the original paper of Carlsson–de Silva is a very readable introduction to the theory, and Oudot's book [124] gives a more detailed treatment. See [18] for results of Botnan-Lesnick on stability of zigzag persistence.

Chapter 8
Multiparameter Persistent Homology

In [33], Carlsson–Zomorodian introduced an extension of persistent homology to the setting of filtrations depending on more than one parameter. Near the end of their paper, they write

> "Our study of multigraded objects shows that no complete discrete invariant exists for multidimensional persistence. We still desire a discriminating invariant that captures persistent information, that is, homology classes with large persistence."

In this chapter we use tools of multigraded algebra to explore *multiparameter persistent homology* (MPH). While there are natural analogs of the free summands (long bars) and torsion components (short bars) of persistent homology, a new phenomenon arises. In particular, there are intermediate components, which are not full dimensional (not long bars), but do not die in high degree (not short bars).

In Sect. 8.3 of this chapter, we introduce the *associated primes* of a module, which provide a useful tool for analyzing multiparameter persistent homology. Just as persistent homology is a $\mathbb{Z}$-graded module over $\mathbb{K}[x]$, an MPH module with n-parameters is a $\mathbb{Z}^n$-graded module M over $\mathbb{K}[x_1, \ldots, x_n]$. This strongly constrains the associated primes of M, and implies that M has an interpretation in terms of translates of coordinate subspaces. This chapter covers

- Definition and Examples.
- Graded algebra, Hilbert function, series, polynomial.
- Associated Primes and $\mathbb{Z}^n$-graded modules.
- Filtrations and Ext.

The recent paper of Botnan-Lesnick [19] gives a state of the art survey of MPH.

H. Schenck, *Algebraic Foundations for Applied Topology and Data Analysis*,
Mathematics of Data 1, https://doi.org/10.1007/978-3-031-06664-1_8

8.1   Definition and Examples

In this section we define multiparameter persistent homology; while the definitions
are a bit lengthy the thing to keep in mind is that the constructions are simply the
$\mathbb{Z}^n$-graded analogs of persistent homology. We illustrate this by working through
an example from [33] in detail. The subsequent sections introduce the tools from
multigraded algebra used in [88] to analyze multiparameter persistent homology.

8.1.1   Multiparameter Persistence

This section develops MPH, following [33].

Definition 8.1.1 Denote by $\mathbb{Z}^n$ the set of n-tuples of integers, and define the
following partial order on $\mathbb{Z}^n$: for any pair of elements $\mathbf{u}, \mathbf{v} \in \mathbb{Z}^n$ we define $\mathbf{u} \preceq \mathbf{v}$ iff
$u_i \leq v_i$ for all $i = 1, \ldots n$, where we write $\mathbf{u} = (u_1, \ldots, u_n)$ and $\mathbf{v} = (v_1, \ldots, v_n)$.
Given a collection of simplicial complexes $\{\Delta_\mathbf{u}\}_{\mathbf{u}\in\mathbb{Z}^n}$ indexed by $\mathbb{Z}^n$, we say that
$\{\Delta_\mathbf{u}\}_{\mathbf{u}\in\mathbb{Z}^n}$ is an *n-filtration* if whenever $\mathbf{u} \preceq \mathbf{v}$ we have that $\Delta_\mathbf{u} \subseteq \Delta_\mathbf{v}$. If there
exists $\mathbf{u}' \in \mathbb{Z}^n$ such that $\Delta_\mathbf{u} = \Delta_{\mathbf{u}'}$ for all $\mathbf{u} \succeq \mathbf{u}'$, then we say that the *n*-filtration
stabilizes. A *multifiltration* is an *n*-filtration for some *n*. It is possible to grade over
other posets, but we shall focus on $\mathbb{Z}^n$.

An *n-filtered simplicial complex* is a simplicial complex Δ together with a
multifiltration $\{\Delta_\mathbf{u}\}_{\mathbf{u}\in\mathbb{Z}^n}$ that stabilizes and such that $\Delta = \cup_{\mathbf{u}\in\mathbb{Z}^n}\Delta_\mathbf{u}$. An *n*-filtered
simplicial complex $(\Delta, \{\Delta_\mathbf{u}\}_{\mathbf{u}\in\mathbb{Z}^n})$ is *finite* if Δ is finite. A *multifiltered simplicial
complex* is an *n*-filtered simplicial complex for some $n \geq 1$.

Given a multifiltered simplicial complex $(\Delta, \{\Delta_\mathbf{u}\}_{\mathbf{u}\in\mathbb{Z}^n})$, for each $x \in \Delta$ we call
the minimal elements $\mathbf{u} \in \mathbb{Z}^n$ (with respect to the partial order $\preceq$) at which it enters
the filtration its *entry degrees*. If every $x \in \Delta$ has exactly one entry degree we call
the multifiltered space *one-critical*.

Let Δ be a multifiltered simplicial complex, and let $i = 0, 1, 2, \ldots$. For any
$\mathbf{u} \in \mathbb{Z}^n$ denote by $C_i(\Delta_\mathbf{u})$ the $\mathbb{K}$-vector space with basis given by the i-simplices
of $\Delta_\mathbf{u}$, and similarly by $H_i(\Delta_\mathbf{u})$ the ith simplicial homology with coefficients in
$\mathbb{K}$. Whenever $\mathbf{u} \preceq \mathbf{v}$ we have that the inclusion maps $\Delta_\mathbf{u} \to \Delta_\mathbf{v}$ induce $\mathbb{K}$-linear
maps $\psi_{\mathbf{u},\mathbf{v}}: C_i(\Delta_\mathbf{u}) \to C_i(\Delta_\mathbf{v})$ and $\phi_{\mathbf{u},\mathbf{v}}: H_i(\Delta_\mathbf{u}) \to H_i(\Delta_\mathbf{v})$ such that whenever
$\mathbf{u} \preceq \mathbf{w} \preceq \mathbf{v}$ we have that $\psi_{\mathbf{w},\mathbf{v}} \circ \psi_{\mathbf{u},\mathbf{w}} = \psi_{\mathbf{u},\mathbf{v}}$, and similarly $\phi_{\mathbf{w},\mathbf{v}} \circ \phi_{\mathbf{u},\mathbf{w}} = \phi_{\mathbf{u},\mathbf{v}}$. We
thus give the following definition:

Definition 8.1.2 Let Δ be a multifiltered simplicial complex. The ith-*chain module*
of Δ over $\mathbb{K}$ is the tuple

$$\left(\{C_i(\Delta_\mathbf{u})\}_{\mathbf{u}\in\mathbb{Z}^n}, \{\psi_{\mathbf{u},\mathbf{v}}: C_i(\Delta_\mathbf{u}) \to C_i(\Delta_\mathbf{v})\}_{\mathbf{u}\preceq\mathbf{v}}\right).$$

Similarly, the *simplicial homology* with coefficients in $\mathbb{K}$ of Δ is the tuple

$$\left(\{H_i(\Delta_{\mathbf{u}})\}_{\mathbf{u}\in\mathbb{Z}^n}, \{\phi_{\mathbf{u},\mathbf{v}} : H_i(\Delta_{\mathbf{u}}) \to H_i(\Delta_{\mathbf{v}})\}_{\mathbf{u}\preceq\mathbf{v}}\right),$$

where the maps $\psi_{\mathbf{u},\mathbf{v}}$ and $\phi_{\mathbf{u},\mathbf{v}}$ are those induced by the inclusions.

Definition 8.1.3 An n-parameter persistence module (or *multiparameter persistence module*) is defined by an n-tuple $\left(\{M_{\mathbf{u}}\}_{\mathbf{u}\in\mathbb{Z}^n}, \{\phi_{\mathbf{u},\mathbf{v}} : M_{\mathbf{u}} \to M_{\mathbf{v}}\}_{\mathbf{u}\preceq\mathbf{v}}\right)$ where $M_{\mathbf{u}}$ is a $\mathbb{K}$-module for each $\mathbf{u}$ and $\phi_{\mathbf{u},\mathbf{v}}$ is a $\mathbb{K}$-linear map, such that whenever $\mathbf{u} \preceq \mathbf{w} \preceq \mathbf{v}$ we have $\phi_{\mathbf{w},\mathbf{v}} \circ \phi_{\mathbf{u},\mathbf{w}} = \phi_{\mathbf{u},\mathbf{v}}$. A *morphism* of multiparameter persistence modules

$$f : \left(\{M_{\mathbf{u}}\}_{\mathbf{u}\in\mathbb{Z}^n}, \{\phi_{\mathbf{u},\mathbf{v}}\}_{\mathbf{u}\preceq\mathbf{v}}\right) \to \left(\{M'_{\mathbf{u}}\}_{\mathbf{u}\in\mathbb{Z}^n}, \{\phi'_{\mathbf{u},\mathbf{v}}\}_{\mathbf{u}\preceq\mathbf{v}}\right)$$

is a collection of $\mathbb{K}$-linear maps $\{f_{\mathbf{u}} : M_{\mathbf{u}} \to M'_{\mathbf{u}}\}_{\mathbf{u}\in\mathbb{Z}^n}$ such that $f_{\mathbf{v}} \circ \phi_{\mathbf{u},\mathbf{v}} = \phi'_{\mathbf{u},\mathbf{v}} \circ f_{\mathbf{u}}$ for all $\mathbf{u} \preceq \mathbf{v}$.

Let Δ be a multifiltered simplicial complex. An example of a morphism of MPH modules is given by the differentials of the simplicial chain complex

$$C_\bullet(\Delta_{\mathbf{u}}) : C_i(\Delta_{\mathbf{u}}) \xrightarrow{d_i} C_{i-1}(\Delta_{\mathbf{u}}),$$

for each $\mathbf{u} \in \mathbb{Z}^n$, which induce morphisms of multiparameter persistence modules

$$\left(\{C_i(\Delta_{\mathbf{u}})\}_{\mathbf{u}\in\mathbb{Z}^n}, \{\psi_{\mathbf{u},\mathbf{v}}\}_{\mathbf{u}\preceq\mathbf{v}}\right) \to \left(\{C_{i-1}(\Delta_{\mathbf{u}})\}_{\mathbf{u}\in\mathbb{Z}^n}, \{\psi_{\mathbf{u},\mathbf{v}}\}_{\mathbf{u}\preceq\mathbf{v}}\right) \tag{8.1.1}$$

for any $i \geq 0$ and where $C_{-1}(\Delta_{\mathbf{u}}) = 0$. Notice that in the spirit of Chap. 7 it is also possible to define an n-parameter persistence module as a functor from the poset category $\mathbb{Z}^n$ to the category with objects $\mathbb{K}$-vector spaces and morphisms $\mathbb{K}$-linear maps. Indeed, in [33] Carlsson–Zomorodian show the categories of MPH modules and $\mathbb{Z}^n$-graded modules are isomorphic.

Example 8.1.4 In [29], Carlsson–Singh–Zomorodian analyze the simplicial homology of the one-critical bifiltration in Fig. 8.1 on the next page.

The differentials in the multifiltered simplicial chain complex are given by

$$d_1 = \begin{bmatrix} -x_2 & -x_2 & 0 & 0 & 0 & 0 & 0 & 0 \\ x_1 x_2^2 & 0 & -x_1^2 x_2^2 & -x_2^2 & 0 & 0 & 0 & 0 \\ 0 & 0 & x_1^2 x_2^2 & 0 & -x_1 & -x_2 & 0 & 0 \\ 0 & 0 & 0 & 0 & 1 & 0 & -1 & 0 \\ 0 & 0 & 0 & 0 & 0 & x_2 & x_1 & -x_1^2 \\ 0 & x_1 x_2^2 & 0 & x_2^2 & 0 & 0 & 0 & x_1^2 \end{bmatrix}$$

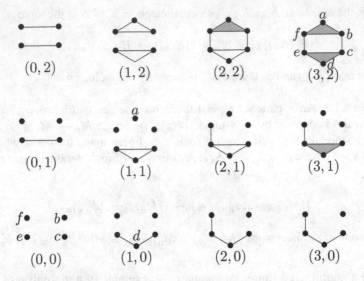

Fig. 8.1 A one-critical bifiltration, given as example in [29]

and

$$d_2 = \begin{bmatrix} x_1 & 0 \\ -x_1 & 0 \\ 0 & 0 \\ x_1^2 & 0 \\ 0 & x_1^2 x_2 \\ 0 & -x_1^3 \\ 0 & x_1^2 x_2 \\ 0 & 0 \end{bmatrix},$$

where the bases of 0, 1 and 2-simplices are ordered lexicographically. For example, the first column of d_1 reflects the fact that the edge $[ab]$ appears at position $(1, 2)$, so has bidegree $x_1 x_2^2$, while $[a]$ appears in position $(1, 1)$ so has bidegree $x_1 x_2$, and $[b]$ appears in position $(0, 0)$. To preserve the *grading*, we need to decorate the usual boundary $\partial_1([ab]) = [b] - [a]$ with coefficients reflecting degree:

$$d_1([ab]) = x_1 x_2^2 [b] - x_2 [a].$$

An easy calculation shows that $\ker(d_1)$ is generated by the rows of

$$L = \begin{bmatrix} 0 & 0 & 0 & 0 & x_2 & -x_1 & x_2 & 0 \\ -1 & 1 & 0 & -x_1 & 0 & 0 & 0 & 0 \\ 0 & 0 & 1 & -x_1^2 & x_1 x_2^2 & 0 & x_1 x_2^2 & x_2^2 \end{bmatrix},$$

There are two obvious relations

$$x_1^2 \cdot \text{row}_1(L) = \text{col}_2(d_2)$$
$$-x_1 \cdot \text{row}_2(L) = \text{col}_1(d_2)$$

These are the only relations, so $H_1(\Delta)$ is the cokernel of the map below:

$$R(-3,-1) \oplus R(-2,-2) \xrightarrow{\begin{bmatrix} x_1^2 & 0 \\ 0 & -x_1 \\ 0 & 0 \end{bmatrix}} R(-1,-1) \oplus R(-1,-2) \oplus R(-2,-2).$$

The shifts in the terms $R(*, *)$ reflect the grading on $R = \mathbb{K}[x_1, x_2]$, which we discuss in detail in the next section. For now, note that the columns of d_1 are graded by the degrees of the edges: for example the last column corresponds to edge $[ef]$, which appears in degree $(2, 0)$. The last row of L corresponds to a generator for H_1, and when multiplied against d_1 has all entries of degree $(2, 2)$.

8.2 Graded Algebra, Hilbert Function, Series, Polynomial

In certain situations, graded algebra turns complicated questions about algebraic objects into questions about vector spaces. This section covers the main invariants of graded modules: the Hilbert function and the Hilbert series, as well as how to read off a first measure of the size of a module–the *rank*. For a module M over a PID, the rank of M is the rank of the free summand, so corresponds in TDA to the number of long bars. In particular, the rank of an MPH module is a natural choice of proxy for the number of long bars.

Definition 8.2.1 Let R be an integral domain, $\mathbb{K} = R_{(0)}$ the field of fractions as in Exercise 2.3.2, and M a finitely generated R-module. The *rank* of M is

$$\text{rank}(M) = \dim_{\mathbb{K}}(M \otimes_R \mathbb{K}) = \dim_{\mathbb{K}}(M_{(0)})$$

Example 8.2.2 The $\mathbb{Z}$-module $\mathbb{Z}^m \bigoplus_{i=1}^{n} \mathbb{Z}/q_i$ has rank m.

Definition 8.2.3 Let G be an abelian group. A ring R is G-graded if

$$R = \bigoplus_{g \in G} R_g,$$

where if $r_i \in R_i$ and $r_j \in R_j$ then

$$r_i \cdot r_j \in R_{i+j}.$$

Note that each R_i is itself a module over R_0. A graded module M over R is a module with a direct sum decomposition

$$M = \bigoplus_{g \in G} M_g,$$

where if $r_i \in R_i$ and $m_j \in M_j$ then

$$r_i \cdot m_j \in M_{i+j}.$$

For applications in topological data analysis, the case of interest is $G = \mathbb{Z}^n$. For these modules, we can determine the rank of a module from the growth rate of the Hilbert Function.

Example 8.2.4 A polynomial ring $R = \mathbb{K}[x_1, \ldots, x_n]$ is $\mathbb{Z}$-graded. The component R_i is the $R_0 = \mathbb{K}$-vector space of polynomials $f(x_1, \ldots, x_n)$ such that every monomial appearing f is of the same degree i, as in Exercise 2.4.16. A simple induction on the number of variables shows that

$$\dim_{\mathbb{K}} R_i = \binom{n-1+i}{i}.$$

8.2.1 The Hilbert Function

For a G-graded module M over a G-graded ring R with $R_0 = \mathbb{K}$ a field, the Hilbert function of M records the dimensions of the vector spaces M_g, $g \in G$. Henceforth, we restrict to the case $G = \mathbb{Z}^n$.

Definition 8.2.5 Let M be a $\mathbb{Z}^n$-graded module over $R = \mathbb{K}[x_1, \ldots, x_n]$. The *Hilbert function* of M is the function $HF(M, \mathbf{u}) = \dim_{\mathbb{K}} M_{\mathbf{u}}$.

Example 8.2.6 For the polynomial ring $R = \mathbb{K}[x_1, \ldots, x_n]$, the two most common gradings are the $\mathbb{Z}$-grading of Example 8.2.4, and the *fine* or $\mathbb{Z}^n$-grading. In the fine grading with $\mathbf{u} = (u_1, \ldots, u_n)$, $R_{\mathbf{u}}$ is one dimensional with basis $x^{\mathbf{u}}$. For $1 \leq r \leq n$ there are other $\mathbb{Z}^r$-gradings possible. For example, consider the ring $T = \mathbb{K}[x_1, \ldots, x_4]$, where $\{x_1, x_2\}$ have degree $(1, 0)$ and $\{x_3, x_4\}$ have degree $(0, 1)$. This is a $\mathbb{Z}^2$-grading (or *bigrading*). The bidegree $(1, 2)$ component of T is

$$T_{(1,2)} = \mathrm{Span}\{x_1 x_3^2, x_1 x_3 x_4, x_1 x_4^2, x_2 x_3^2, x_2 x_3 x_4, x_2 x_4^2\}$$

A $\mathbb{Z}^n$-graded Hilbert function is an n-dimensional array, with the $\mathbf{u} = (u_1, \ldots, u_n)$ entry equal to the dimension of $M_{\mathbf{u}}$. In the fine grading, the Hilbert function of R itself is an n-dimensional array with a one in every position which has all indices non-negative, and zeroes elsewhere. The Hilbert function of $R(-\mathbf{u})$ is the same array, but with the origin translated to position $\mathbf{u}$; this process is known as *shifting*.

Definition 8.2.7 For a finitely generated R-module M, a *free resolution* is an exact sequence of free modules terminating in M:

$$\cdots \xrightarrow{d_3} F_2 \xrightarrow{d_2} F_1 \xrightarrow{d_1} F_0 \xrightarrow{d_0} M \longrightarrow 0.$$

When working in the category of graded rings and graded modules, maps must preserve the graded structure: $\phi : M \to N$ takes $M_{\mathbf{u}} \to N_{\mathbf{u}}$.

Example 8.2.8 In Example 2.4.17 we considered the ideal I of 2×2 minors of a 2×3 matrix $A = \begin{bmatrix} x & y & z \\ y & z & w \end{bmatrix}$. If we grade R by $\mathbb{Z}$, then a free $\mathbb{Z}$-graded resolution for $M = R/I$ is given by

$$0 \longrightarrow R(-3)^2 \xrightarrow{\begin{bmatrix} x & y \\ -y & -z \\ z & w \end{bmatrix}} R(-2)^3 \xrightarrow{\begin{bmatrix} xz - y^2 & xw - yz & yw - z^2 \end{bmatrix}} R \longrightarrow R/I$$

Elements of the kernel of d_i are called i^{th} *syzygies*. In the special situation above, the first syzygies arise by stacking $[x, y, z]$ or $[y, z, w]$ as a third row of A; the determinant of the resulting 3×3 matrix has a repeat row, and hence yields a syzygy. The leftmost matrix has no kernel so the process stops. Since the map

$$F_1 = R^3 \xrightarrow{d_1} R^1 = F_0$$

is multiplication by quadrics, to preserve the $\mathbb{Z}$-grading, we must shift so that the degrees of the generators of F_1 have degree 2. Similarly, as the leftmost map d_1 has degree one entries and maps onto a target generated in degree two, the generators of F_2 are in degree three.

In nice situations, free resolutions are actually finite; for a proof see [70].

Theorem 8.2.9 (Hilbert Syzygy Theorem) *A finitely generated $R = \mathbb{K}[x_1, \ldots, x_n]$ module M has a free resolution of length at most n.*

Lemma 8.2.10 *For a finitely generated graded* $R = \mathbb{K}[x_1, \ldots, x_n]$-*module* M *with free resolution* $F_{\bullet}$ *as in Definition 8.2.7,*

(a) $HF(M, \mathbf{u}) = \sum_{i \geq 0} (-1)^i HF(F_i, \mathbf{u})$

(b) $\mathrm{rank}_R(M) = \sum_{i \geq 0} (-1)^i \mathrm{rank}_R(F_i).$

Proof The key point is that because the maps in a graded free resolution preserve the grading, the free resolution is exact in each graded degree. Since for each degree $\mathbf{u}$, the degree $\mathbf{u}$ components of $F_{\bullet}$ are an exact sequence of vector spaces over $\mathbb{K}$, so (a) follows from Example 4.2.4. For part (b), by Theorem 8.2.9 M has a finite free resolution. Tensoring the (finite) exact sequence of Definition 8.2.7 with the field of fractions of R is equivalent to localizing. By Theorem 2.3.3 localization is exact, so (b) follows from Example 4.2.4.   □

Lemma 8.2.11 *For a finitely generated* $\mathbb{Z}^n$-*graded* $R = \mathbb{K}[x_1, \ldots, x_n]$-*module* M,

$$\mathrm{rank}_R(M) = HF(M, \mathbf{u}) \text{ for } \mathbf{u} \gg \mathbf{0}.$$

Proof A finitely generated, $\mathbb{Z}^n$-graded free module F is of the form

$$F = \bigoplus_{j=1}^{m} R(-\mathbf{u}_j), \text{ with } \mathbf{u}_j \in \mathbb{Z}^n.$$

Since $HF(R(-\mathbf{u}_j), \mathbf{v}) = 1$ if $\mathbf{v} \geq \mathbf{u}_j$ and 0 otherwise, when $\mathbf{v} \geq \mathbf{u}_j$ for all $j \in \{1, \ldots, n\}$

$$HF(F, \mathbf{v}) = n = \mathrm{rank}_R(F).$$

By Theorem 8.2.9 M has a finite free resolution; apply Lemma 8.2.10.   □

Example 8.2.12 (Hilbert Polynomial) In the special case of a $\mathbb{Z}$-graded ring R as in Example 8.2.8, the Hilbert function of a graded R-module M is given by a polynomial when $i \in \mathbb{Z}$ is sufficiently large:

$$HF(M, i) = P(M, i) \in \mathbb{Q}[i], \text{ when } i \gg 0.$$

This can be proved by induction on the number of variables, using the four term exact sequence

$$0 \longrightarrow \ker(\cdot x_n) \longrightarrow M(-1) \overset{\cdot x_n}{\longrightarrow} M \longrightarrow \mathrm{coker}(\cdot x_n) \longrightarrow 0,$$

using that $\ker(\cdot x_n)$ and $\mathrm{coker}(\cdot x_n)$ are $\mathbb{K}[x_1, \ldots, x_{n-1}]$-modules. The polynomial $P(M, i)$ has the form

$$P(M, i) = \frac{a_m}{m!} i^m + \ldots + a_0, \text{ with } m \leq n - 1.$$

In the context of projective geometry with $M = R/I$ as in Example 8.2.8, the *dimension* of $\mathbf{V}(I)$ is the degree of $P(R/I, i)$, and the *degree* of $\mathbf{V}(I)$ is the (normalized) lead coefficient a_m. The dimension of $\mathbf{V}(I)$ is the number of times it can be cut with a general hyperplane until we're left with a set of points; the number of points (with multiplicity, if $\mathbf{V}(I)$ is singular) is the degree of $\mathbf{V}(I)$. By Example 8.2.4

$$P(R(-j), i) = \binom{n - 1 + i - j}{n - 1}$$

Exercise 8.2.13 Use the free resolution which appears in Example 8.2.8 to show that $P(R/I, i) = 3i + 1$. Thus the degree of $\mathbf{V}(I)$ is three, and the dimension is one. The corresponding curve in $\mathbb{P}^3$ is known as the *twisted cubic*. ◇

8.2.2 The Hilbert Series

The Hilbert series provides a compact way of packaging the data of the Hilbert function into a formal power series.

Definition 8.2.14 For a finitely generated $R = \mathbb{K}[x_1, \ldots, x_n]$-module M which is $\mathbb{Z}^n$-graded, the *multigraded Hilbert series* of M is the formal power series in $\mathbb{Z}[[t_1, \ldots, t_n]]$ defined as follows:

$$HS(M, \mathbf{t}) = \sum_{\mathbf{u} \in \mathbb{Z}^n} HF(M, \mathbf{u}) \mathbf{t}^{\mathbf{u}}.$$

Inducting on the number of variables yields

$$HS(R(-\mathbf{u}), \mathbf{t}) = \frac{\mathbf{t}^{\mathbf{u}}}{\prod_{i=1}^n (1 - t_i)}. \tag{8.2.1}$$

By Lemma 8.2.10 the Hilbert function is additive on exact sequences, so this is also true for the Hilbert series. Hence applying Theorem 8.2.9 we have

$$HS(M, \mathbf{t}) = \sum_{i=0}^n (-1)^i HS(F_i, \mathbf{t}). \tag{8.2.2}$$

For a finitely generated, multigraded R-module M, it follows from the Hilbert Syzygy Theorem and Eqs. 8.2.1–8.2.2 that the multigraded Hilbert series of M is a rational polynomial of the form

$$HS(M, \mathbf{t}) = \frac{P(t_1, \ldots, t_n)}{\prod_{i=1}^{n}(1 - t_i)}. \tag{8.2.3}$$

The polynomial $P(t_1, \ldots, t_n)$ in Eq. 8.2.3 is an invariant of the module.

Lemma 8.2.15 *For a finitely generated graded* $R = \mathbb{K}[x_1, \ldots, x_n]$*-module* M, $\mathrm{rank}_R(M)$ *is equal to* $P(\mathbf{1})$, *where* P *is as in Equation 8.2.3.*

Proof To see that $\mathrm{rank}_R(M)$ is the numerator of $HS(M, \mathbf{t})$ evaluated at $\mathbf{1}$, note that it holds for free modules by Eq. 8.2.1. Now apply Lemma 8.2.10.   □

Example 8.2.16 Note that if M is a one-parameter persistence module, we have

$$HS(M, t) = \frac{\sum_{\text{long bars}} t^{b_l}}{(1 - t)} + \frac{\sum_{\text{short bars}}(t^{b_s} - t^{d_s})}{(1 - t)} \tag{8.2.4}$$

where b_l is the birth time for a long bar, b_s the birth time for a short bar, and d_s the death time of a short bar. So

$$HS(M, 1) = \mathrm{rank}_R(M)$$

is indeed the number of long bars.

We close by revisiting Example 8.1.4

Example 8.2.17 In Example 8.1.4, $H_1(\Delta) = \mathrm{coker}(\delta_1)$, with

$$\delta_1 = \begin{bmatrix} x_1^2 & 0 \\ 0 & -x_1 \\ 0 & 0 \end{bmatrix}$$

It is easy to see that δ_1 has no kernel itself, so we have a free resolution for $H_1(\Delta)$

$$0 \longrightarrow F_1 \xrightarrow{\delta_1} F_0 \longrightarrow H_1(\Delta) \longrightarrow 0,$$

so

$$
\begin{aligned}
HS(H_1(\Delta), \mathbf{t}) &= HS(F_0, \mathbf{t}) - HS(F_1, \mathbf{t}) \\
&= \frac{t_1 t_2 + t_1 t_2^2 + t_1^2 t_2^2}{(1 - t_1)(1 - t_2)} - \frac{t_1^3 t_2 + t_1^2 t_2^2}{(1 - t_1)(1 - t_2)} \\
&= \frac{t_1 t_2 + t_1 t_2^2 - t_1^3 t_2}{(1 - t_1)(1 - t_2)}.
\end{aligned}
$$

It follows that $P(H_1(\Delta), \mathbf{1}) = 1$, so $\text{rank}(H_1(\Delta)) = 1$. In fact, because in this case the presentation is so simple, we see that

$$H_1(\Delta) \simeq (R(-1, -1)/\langle x_1^2 \rangle) \bigoplus (R(-1, -2)/\langle x_1 \rangle) \bigoplus R(-2, -2).$$

In particular, note that the first two terms

$$(R(-1, -1)/\langle x_1^2 \rangle) \bigoplus (R(-1, -2)/\langle x_1 \rangle)$$

are not free, but also do not vanish in high degree in the x_2 variable. For any principal ideal $\langle f \rangle$ with $\deg(f) = \mathbf{u}$, we have an exact sequence:

$$0 \longrightarrow R(-\mathbf{u}) \overset{\cdot f}{\longrightarrow} R \longrightarrow R/\langle f \rangle \longrightarrow 0,$$

so

$$HS(R/\langle f \rangle, \mathbf{t}) = \frac{1 - t^{\mathbf{u}}}{\prod_{i=1}^n (1 - t_i)}.$$

The next section introduces *associated primes*, which will provide a way to understand the structure of MPH modules, which generally do not decompose as a direct sum as they do in this example. As a module, the summand $R(-1, -2)/\langle x_1 \rangle$ above is supported in the sense of Definition 8.3.4 on $\mathbf{V}(x_1)$.

Exercise 8.2.18 Show $HS(R(-1, -2)/\langle x_1 \rangle, \mathbf{t}) = \frac{t_1 t_2^2}{1 - t_2}$. ◇

8.3 Associated Primes and $\mathbb{Z}^n$-Graded Modules

In this section, we will use the extra structure provided by the $\mathbb{Z}^n$-grading to analyze MPH modules. As a first question, how can we visualize a module? Sheaf theory provides the right tools to tackle this question.

8.3.1 Geometry of Sheaves

We saw in Chap. 3 that vector bundles are topological spaces that can be visualized (locally) in a natural way. Is it possible to visualize a sheaf? The answer is yes, and the intuition comes from vector bundles. A rank one vector bundle on a curve might look like

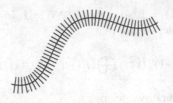

The gluing axioms for sheaves satisfy the same properties required to define transition functions for vector bundles. For vector bundles the base space B may be complicated, but the fibers are all simply vector spaces. For sheaves on varieties, our focus will be when the base space is an affine variety as in Definition 3.3.6, or on sheaves on $\mathbb{A}^n_{\mathbb{K}}$ itself.

Definition 8.3.1 As a set, an affine variety is simply $\mathbf{V}(I)$. Since $\mathbf{V}(I) \subseteq \mathbb{A}^n_{\mathbb{K}}$, it has the induced Zariski topology from $\mathbb{A}^n$. On $\mathbb{A}^n$, regular functions are polynomials in $R = \mathbb{K}[x_1, \ldots, x_n]$, and since $f \in I$ is the zero function on $\mathbf{V}(I)$, regular functions on $\mathbf{V}(I)$ are polynomials defined up to I, that is, elements of R/I.

The fundamental sheaf in algebraic geometry is the sheaf of regular functions.

Definition 8.3.2 On an affine variety $X = \mathbf{V}(I) \subseteq \mathbb{A}^n$ and Zariski open set U, the sheaf of regular functions is defined as

$$\mathscr{O}_X(U) = \left\{ \frac{f}{g} \mid f, g \in R/I \right\},$$

with $g(p) \neq 0$ for $p \in U$; since U is open, $U = \mathbf{V}(J)^c$ for some ideal J. By the Noetherian condition, the sets $U_f = \mathbf{V}(f)^c$ for all $f \in R$ are a basis for the Zariski topology, and any $\mathscr{M}$ which is a sheaf of $\mathscr{O}_X$-modules can be defined by starting with an ordinary R-module M, and defining $\mathscr{M}$ on the basis via

$$\mathscr{M}(U_f) = M_f, \text{ with the stalks corresponding to } \mathscr{M}_p = M_p,$$

where M_f and M_p are obtained from the localization operation of Chap. 2.

Definition 8.3.3 The *tangent sheaf* is constructed as in Example 3.3.4:

$$\mathscr{T}_X(U_i) = \mathrm{Der}_{\mathbb{K}} \mathscr{O}_X(U_i)$$

A point $p \in X$ is *smooth* if the stalk $\mathscr{T}_{X,p}$ is a free $\mathscr{O}_{X,p}$-module; X is smooth if every point is a smooth point. This mirrors the construction of the *tangent bundle* of an n-dimensional *real manifold* M. In this situation, $\mathbb{K} = \mathbb{R}$ and there is a cover of M by open sets U_i diffeomorphic to $\mathbb{R}^n$. The transition maps ϕ_{ij} appearing in Definition 3.2.2 are C^∞ maps between charts, and the fibers are $\mathbb{R}^n$.

Definition 8.3.4 The *support* of an R-module M consists of the prime ideals P such that $M_P \neq 0$.

When M is a finitely generated R-module, a prime ideal P is in the support of M iff $ann(M) \subseteq P$. If M is generated by $\{m_1, \ldots, m_k\}$, then there is a surjection

$$R^k \longrightarrow M \longrightarrow 0.$$

Localizing at a prime P makes elements of R outside P invertible, therefore if $ann(M) \nsubseteq P$, then some element $r \in ann(M)$ is a unit in R_P. But a unit can annihilate every element of M_P only if $M_P = 0$.

8.3.2 Associated Primes and Primary Decomposition

We defined the variety of an ideal $I \subseteq \mathbb{K}[x_1, \ldots, x_n]$ in Chap. 2:

$$\mathbf{V}(I) = \{p \in \mathbb{K}^n \mid f(p) = 0 \text{ for all } f \in I\}$$

Example 8.3.5 For $I = \langle x^2 - x, xy - x \rangle \subset \mathbb{R}[x, y]$, $\mathbf{V}(I) \subseteq \mathbb{R}^2$ is pictured below:

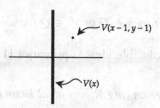

So $\mathbf{V}(I)$ consists of the line $\mathbf{V}(x)$ and the point $\mathbf{V}(x - 1, y - 1)$. Notice that

$$I = \langle x \rangle \cap \langle x - 1, y - 1 \rangle$$

By Theorem 2.2.11, I is prime iff R/I is a domain, and since

$$\mathbb{R}[x, y]/\langle x \rangle \simeq \mathbb{R}[y] \text{ and } \mathbb{R}[x, y]/\langle x - 1, y - 1 \rangle \simeq \mathbb{R}$$

are both domains, we see that I is the intersection of two prime ideals.

It is clearly too much to hope that any ideal is an intersection of prime ideals:

$$\langle x^2 \rangle \subseteq \mathbb{R}[x]$$

cannot be written as an intersection of prime ideals. But it turns out that any ideal in a Noetherian ring has a decomposition as a finite intersection of *irreducible* ideals, which appeared in Definition 2.2.9. Recall an ideal I is irreducible if it is not the intersection of two proper ideals, each strictly containing I.

Theorem 8.3.6 *In a Noetherian ring R, any ideal I can be written as*

$$I = \bigcap_{i=1}^{n} I_i,$$

with I_i irreducible.

Proof Let Σ be the set of all ideals which cannot be written as a finite intersection of irreducible ideals. Since R is Noetherian, Σ has a maximal element I'. Since I' is reducible,

$$I' = J_1 \cap J_2$$

with $I' \subsetneq J_1$ and $I' \subsetneq J_2$. Since I' is maximal, neither J_1 nor J_2 are in Σ. Hence both J_1 and J_2 can be written as finite intersections of irreducibles, thus I' is a finite intersection of irreducibles, a contradiction. $\qquad\qquad\square$

Primary ideals also appeared in Definition 2.2.9; Q is primary if

$$xy \in Q \Rightarrow x \in Q \text{ or } y^n \in Q \text{ for some } n.$$

In a Noetherian ring an irreducible ideal is primary ([133] Lemma 1.3.3), so we have

Theorem 8.3.7 *In a Noetherian ring R, any ideal I can be written as*

$$I = \bigcap_{i=1}^{n} Q_i,$$

with Q_i primary.

It follows immediately from the definition that the radical of a primary ideal is prime, since x or $y^n \in Q \Rightarrow x$ or $y \in \sqrt{Q}$.

Definition 8.3.8 For a primary decomposition

$$I = \bigcap_{i=1}^{n} Q_i$$

the $P_i = \sqrt{Q_i}$ appearing in this decomposition are the *Associated Primes* of I.

Exercise 8.3.9 Show primary decomposition is not unique:

$$\langle xy, x^2 \rangle = \langle x^2, y \rangle \cap \langle x \rangle$$
$$= \langle x^2, xy, y^2 \rangle \cap \langle x \rangle$$

In particular, show $\langle x^2, y \rangle$ and $\langle x^2, xy, y^2 \rangle$ are primary. What are the associated primes?   ◇

There is a version of associated primes for an R-module M:

Definition 8.3.10 A prime ideal P is associated to an R-module M if

$$P = \text{ann}(m) \text{ for some } m \in M,$$

where an element $r \in R$ is in the ideal $\text{ann}(m)$ if $r \cdot m = 0$.

Notice that any $m \in M$ generates a principal submodule $R \cdot m \subseteq M$, so we have a diagram as below, where the vertical arrow is an isomorphism.

$$0 \longrightarrow R \cdot m \longrightarrow M$$

$$\uparrow$$

$$R/ann(m)$$

Exercise 8.3.11 Show that when $M = R/I$ and $I = \cap Q_i$, then for any index i

$$I \subseteq Q_i \subseteq P_i$$

and it is possible to find $m \in M$ so that $ann(m) = P_i$. So in the setting of a cyclic module $M = R/I$, Definitions 8.3.10 and 8.3.8 agree.   ◇

Example 8.3.12 We return to Example 8.3.5 from the beginning of the section, where $R = \mathbb{R}[x, y]$ and $I = \langle x^2 - x, xy - x \rangle$. Let $X = \mathbb{A}^2$, so that R/I defines a sheaf $\mathscr{M}$ of $\mathscr{O}_X$-modules. From the definition of the support of a module, to visualize the stalks $\mathscr{M}_P$, we need to localize at the associated primes.

- If $P = \langle x - 1, y - 1 \rangle$, then $x \notin P$ so x is a unit, and

$$(R/I)_P \simeq R_P/I_P \simeq R_P/\langle x - 1, y - 1 \rangle_P \simeq (\mathbb{R})_P \simeq \mathbb{R}.$$

So the stalk $\mathscr{M}_P$ at the point $(1, 1)$ is just a point.

- If $P = \langle x \rangle$, then $\{x - 1, y - 1\}$ are both units, and

$$(R/I)_P \simeq R_P/I_P \simeq R_P/\langle x \rangle_P \simeq \mathbb{R}[y]_P.$$

So the stalk $\mathscr{M}_P$ along the line $\mathbf{V}(x)$ is a line.

- If P is a prime $\notin \{\langle x \rangle, \langle x - 1, y - 1 \rangle\}$, let P be an ideal of the form

$$\langle x - a, y - b \rangle \text{ with } (a, b) \in \mathbb{R}^2 \setminus \{\mathbf{V}(x), (1, 1)\}.$$

Then $y - 1$ is a unit in R_P, and x is a unit in R_P, so $xy - x \notin P$. Hence $xy - x$ is a unit, which also annihilates $\mathcal{M}$ so $\mathcal{M}_P = 0$.

So $\mathcal{M}_P = 0$ except when $P \in \{\langle x \rangle, \langle x - 1, y - 1 \rangle\}$. We might visualize $\mathcal{M}$ as:

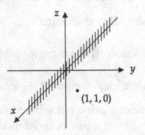

Definition 8.3.13 For a Noetherian ring R, the *dimension* (or Krull dimension) is

$$\dim(R) = \sup_n \{P_0 \subsetneq P_1 \subsetneq \cdots P_n \subsetneq R\} \text{ with the } P_i \text{ prime ideals in } R.$$

If R is an integral domain (which will always be the case for us), there is a unique common starting point for chains as above: $0 = P_0$. Note that if R is not Noetherian, then there can be infinite ascending proper chains of ideals.

Definition 8.3.14 Let R be a ring and I an ideal in R. The codimension of I is

$$\sup_m \{P_0 \subsetneq P_1 \subsetneq P_2 \subsetneq \cdots P_m \subseteq I\}$$

Example 8.3.15 For $R = \mathbb{K}[x_1, \ldots, x_n]$, since

$$\{0 \subseteq \langle x_1 \rangle \subseteq \langle x_1, x_2 \rangle \subseteq \cdots \subseteq \langle x_1, \ldots, x_n \rangle\}$$

is a proper chain of prime ideals, $\dim(R) \geq n$; it is not hard to show this is an equality. Consider the ideal

$$I = \langle x_1, \ldots, x_m \rangle \subseteq R.$$

From the definition, I has codimension at least m, and this too is an equality. Geometric mnemonic: $\mathbf{V}(I) \subseteq \mathbb{K}^n$ is a linear space defined by the vanishing of m independent linear forms, so is of dimension $n - m$, hence of *complimentary* dimension m. So we have

$$\mathrm{codim}(I) = n - \dim(\mathbf{V}(I)).$$

A word of caution: If I is defined by homogeneous polynomials and we work projectively in $\mathbb{P}^{n-1}$, then the dimension of $\mathbf{V}(I)$ as a projective variety is one less than the dimension as a variety in $\mathbb{K}^n$. For example $\mathbf{V}(x_0, x_1)$ defines a plane in $\mathbb{A}^4$ or a line in $\mathbb{P}^3$, and $\langle x_0, x_1 \rangle$ is of codimension two.

8.3.3 Additional Structure in the $\mathbb{Z}^n$-Graded Setting

For $\mathbb{Z}$-graded rings and modules as in Example 8.2.8, we noted that maps also have to preserve gradings; this also applies to $\mathbb{Z}^n$-graded modules.

Theorem 8.3.16 *The only $\mathbb{Z}^n$-graded prime ideals in $R = \mathbb{K}[x_1, \ldots, x_n]$ are*

$$P = \langle x_{i_0}, \ldots, x_{i_r} \rangle. \tag{8.3.1}$$

Proof In the $\mathbb{Z}^n$-grading, the only polynomial of degree $\mathbf{u} = (u_1, \ldots, u_n)$ is

$$x^{\mathbf{u}} = x_1^{u_1} \cdots x_n^{u_n}$$

Hence, the only $\mathbb{Z}^n$-homogeneous ideals are ideals generated by monomials. A prime ideal P cannot contain a monomial of degree two or more as a minimal generator, since if $x_1 x_2 \in P$ then $x_1 \in P$ or $x_2 \in P$. Hence a prime ideal P is generated by monomials iff P is generated by a subset of the variables. $\quad\square$

As an immediate consequence, we have

Corollary 8.3.17 *The associated primes of a fine graded module are subsets of variables. An MPH module is supported on a coordinate subspace arrangement.*

Example 8.3.18 Modify Example 8.1.4 by adding vertex $[g]$, edges $[ag], [bg], [fg]$ and triangles $[abg], [bfg], [afg]$ in degree $(1, 3)$. The corresponding $\mathbb{Z}^2$-filtered complex Δ' is the same as the complex of Example 8.1.4 in degrees $(*, a)$ with $a \leq 2$, and for $a = 3$ is pictured (Fig. 8.2) on the next page.
Keeping the ordering of the faces of Δ from Example 8.1.4, and adding

$[g]$	as the last ordered basis element for	$C_0(\Delta')$
$[ag], [bg], [fg]$	as the last ordered basis elements for	$C_1(\Delta')$
$[abg], [afg], [bfg]$	as the last ordered basis elements for	$C_2(\Delta')$

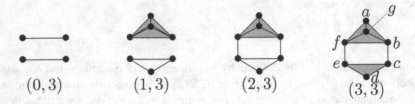

Fig. 8.2 The simplices of degree $(*, 3)$ in Δ'

we have

$$
d_2(\Delta') = \begin{bmatrix}
x_1 & 0 & x_2 & 0 & 0 \\
-x_1 & 0 & 0 & x_2 & 0 \\
0 & 0 & 0 & 0 & 0 \\
x_1^2 & 0 & 0 & 0 & x_1 x_2 \\
0 & x_1^2 x_2 & 0 & 0 & 0 \\
0 & -x_1^3 & 0 & 0 & 0 \\
0 & x_1^2 x_2 & 0 & 0 & 0 \\
0 & 0 & 0 & 0 & 0 \\
0 & 0 & -1 & -1 & 0 \\
0 & 0 & 1 & 0 & -1 \\
0 & 0 & 0 & 1 & 1
\end{bmatrix}
$$

We compute that $\ker(d_2(\Delta'))$ is generated by $[x_2, 0, -x_1, x_1, -x_1]^T$, hence

$$
H_2(\Delta') \simeq R(-1, -3).
$$

Exercise 8.3.19 Compute $d_1(\Delta')$, and show that

$$
H_1(\Delta') \simeq R(-2, -2) \oplus R(-1, -1)/x_1^2 \oplus R(-1, -2)/\langle x_1, x_2 \rangle.
$$

The associated primes are $\{0, \langle x_1 \rangle, \langle x_1, x_2 \rangle\}$. $\diamond$

8.4 Filtrations and Ext

The examples of MPH modules we've encountered so far have all decomposed as direct sums, but this is atypical. In general, our goal will be to stratify or filter the MPH module, to obtain an analog of the barcode. We will use the associated primes to do this, which leads to the first question: in general, how do we identify elements of $\mathrm{Ass}(M)$? There are general algorithms for identifying the associated primes of a $\mathbb{Z}$-graded $R = \mathbb{K}[x_1, \ldots, x_n]$-module. However, these algorithms rely on the use of Gröbner bases [47], and in the worst case Gröbner bases can require doubly exponential runtime. In the setting of multiparameter persistence, Corollary 8.3.17 shows that an associated prime P of M is a subset of the variables. This places *extremely* strong constraints on the structure of fine graded modules; in particular, there is a *finite* subset of possible associated primes.

The algorithm mentioned above for identifying associated primes uses the derived functor Ext, so this is the right time to skip to Chap. 9 and read Sects. 9.1 and 9.2 if you're encountering Ext for the first time. Combining the abstract algebra analysis using Ext with the strong constraints on the associated primes of a fine graded module will provide insight into the structure of MPH, as well as yielding one candidate for a higher dimensional proxy for the barcode.

Theorem 8.4.1 *For a finitely generated $\mathbb{Z}$-graded $R = \mathbb{K}[x_1 \ldots, x_n]$-module M, a prime ideal P of codimension c is in $\mathrm{Ass}(M)$ iff it is in $\mathrm{Ass}(Ext^c(M, R))$.*

Proof See [71] for complete details, or [133] for a quick sketch. □

Example 8.4.2 Let $R = \mathbb{K}[x_1, x_2]$ and $M = R/I$ with $I = \langle x_1^2, x_1 x_2 \rangle$. In Exercise 8.3.9, we saw that the associated primes of I are $\{\langle x_1 \rangle, \langle x_1, x_2 \rangle\}$, which followed from the primary decomposition

$$I = \langle x_1^2, x_2 \rangle \cap \langle x_1 \rangle.$$

Chapter 9 gives a recipe to compute $Ext^i(M, R)$: take a free resolution $F_\bullet$ for M, drop M from the sequence, apply $Hom_R(\bullet, R)$ to $F_\bullet$, and compute homology. This seems like an arduous task, until we roll up our sleeves and do it.

- Step 1: A free resolution for R/I is easy to do by hand:

$$0 \longrightarrow R \xrightarrow{\begin{bmatrix} x_2 \\ -x_1 \end{bmatrix}} R^2 \xrightarrow{\begin{bmatrix} x_1^2 & x_1 x_2 \end{bmatrix}} R \longrightarrow R/I$$

- Step 2: dropping R/I and applying $Hom_R(\bullet, R)$–which simply transposes the differentials–yields

$$0 \longrightarrow R \xrightarrow{\begin{bmatrix} x_1^2 \\ x_1 x_2 \end{bmatrix}} R^2 \xrightarrow{\begin{bmatrix} x_2 & -x_1 \end{bmatrix}} R \longrightarrow 0$$

- Step 3: Compute the homology. Since the leftmost map has no kernel, we see $Ext^0(R/I, R) = 0$. The kernel of the rightmost map is R; since we are computing homology we must quotient, and so

$$Ext^2(R/I, R) \simeq R/\langle x_2, x_1 \rangle \simeq \mathbb{K}.$$

Note the only associated prime of Ext^2 is $\langle x_1, x_2 \rangle$. Finally, we compute

$$\mathrm{Ext}^1(R/I, R) = \ker\left([x_2, -x_1]\right) \Big/ \mathrm{im}\left(\begin{bmatrix} x_1^2 \\ x_1 x_2 \end{bmatrix}\right) = \begin{bmatrix} x_1 \\ x_2 \end{bmatrix} \Big/ \begin{bmatrix} x_1^2 \\ x_1 x_2 \end{bmatrix}$$

The last module has a single generator e_1, and the single relation is $x_1 \cdot e_1$, so we conclude that

$$\mathrm{Ext}^1(R/I, R) \simeq R/\langle x_1 \rangle.$$

In particular, the only associated prime of $\mathrm{Ext}^1(R/I, R)$ is $\langle x_1 \rangle$, illustrating Theorem 8.4.1. For another example of this, see Chap. 9, Example 9.2.8.

Our next example brings the fine grading into the picture.

Example 8.4.3 Consider a $\mathbb{Z}^3$-filtration Δ on vertex set $\{a, b, c, d\}$ below, with all vertices appearing in degree 000, and

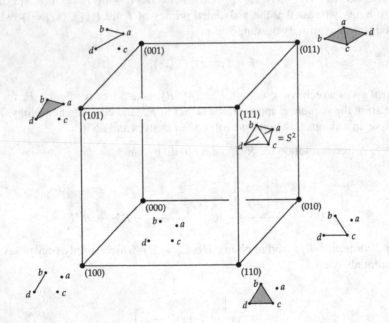

with edges in degrees

$$ab = 001 \quad ac = 011 \quad ad = 001 \quad bc = 010 \quad bd = 100 \quad cd = 010$$

and triangles in degrees

$$abc = 011 \quad abd = 101 \quad acd = 011 \quad bcd = 110.$$

In degree (111) the triangles bound a hollow tetrahedron, so $H_2(\Delta_{111}) \neq 0$.

To see this algebraically, with respect to the ordered bases above we have

$$d_2(\Delta) = \begin{bmatrix} x_2 & x_1 & 0 & 0 \\ -1 & 0 & 1 & 0 \\ 0 & -x_1 & -x_2 & 0 \\ x_3 & 0 & 0 & x_1 \\ 0 & x_3 & 0 & -x_2 \\ 0 & 0 & x_3 & x_1 \end{bmatrix}$$

and

$$d_1(\Delta) = \begin{bmatrix} -x_3 & -x_2 x_3 & -x_3 & 0 & 0 & 0 \\ x_3 & 0 & 0 & -x_2 & -x_1 & 0 \\ 0 & x_2 x_3 & 0 & x_2 & 0 & -x_2 \\ 0 & 0 & x_3 & 0 & x_1 & x_2 \end{bmatrix}$$

Exercise 8.4.4 Show that $H_2(\Delta) \simeq R(1, 1, 1)$ and $H_1(\Delta) = 0$. ◇

The second column of d_1 is a combination of the first and fourth columns so can be dropped. A minimal presentation matrix for $H_0(\Delta)$ is given by

$$\phi = \begin{bmatrix} 0 & 0 & 0 & -x_3 & -x_3 \\ -x_1 & -x_2 & 0 & x_3 & 0 \\ 0 & x_2 & -x_2 & 0 & 0 \\ x_1 & 0 & x_2 & 0 & x_3 \end{bmatrix}$$

A minimal free resolution for $H_0(\Delta)$ is given (omitting gradings) by

$$0 \longrightarrow R^1 \xrightarrow{\begin{bmatrix} x_3 \\ -x_2 \\ x_1 \end{bmatrix}} R^3 \xrightarrow{\begin{bmatrix} -x_2 & -x_3 & 0 \\ x_1 & 0 & -x_3 \\ x_1 & 0 & -x_3 \\ 0 & -x_1 & -x_2 \\ 0 & x_1 & x_2 \end{bmatrix}} R^5 \xrightarrow{\phi} R^4 \longrightarrow H_0(\Delta) \longrightarrow 0.$$

Exercise 8.4.5 Use the free resolution above to prove

$$
\begin{aligned}
Ext^3(H_0(\Delta), R) &\simeq & \mathbb{K} \\
Ext^2(H_0(\Delta), R) &\simeq & 0 \\
Ext^1(H_0(\Delta), R) &\simeq & R/\langle x_2 \rangle \oplus R/\langle x_3 \rangle \\
Hom(H_0(\Delta), R) &\simeq & R^1
\end{aligned}
$$

To do this, we transpose the differentials in the free resolution for $H_0(\Delta)$ and compute homology. For example, to see that $Ext^3(H_0(\Delta), R) \simeq \mathbb{K}$, just note that this module is the cokernel of transpose of the last map above

$$Ext^3(H_0(\Delta), R) = R^1/\operatorname{im}([x_3, -x_2, x_1])$$

which is clearly $\mathbb{K}$.   ◇

It will be useful to display the R-modules $\operatorname{Ext}^i(H_j(\Delta), R)$ in an $(i + 1) \times (j + 1)$ array as below

$$
\begin{array}{ccc}
\operatorname{Ext}^3(H_0(\Delta), R) & \operatorname{Ext}^3(H_1(\Delta), R) & \operatorname{Ext}^3(H_2(\Delta), R) \\
\operatorname{Ext}^2(H_0(\Delta), R) & \operatorname{Ext}^2(H_1(\Delta), R) & \operatorname{Ext}^2(H_2(\Delta), R) \\
\operatorname{Ext}^1(H_0(\Delta), R) & \operatorname{Ext}^1(H_1(\Delta), R) & \operatorname{Ext}^1(H_2(\Delta), R) \\
\operatorname{Hom}(H_0(\Delta), R) & \operatorname{Hom}(H_1(\Delta), R) & \operatorname{Hom}(H_2(\Delta), R)
\end{array}
$$

Since we have shown that $H_2(\Delta)$ is free rank one, for Example 8.4.3 this array takes the form

$$
\begin{array}{ccc}
\mathbb{K} & 0 & 0 \\
0 & 0 & 0 \\
Ext^1(H_1(\Delta), R) & 0 & 0 \\
R^1 & 0 & R^1
\end{array}
$$

In Chap. 9, we'll see that the different spots in this diagram are connected by a *spectral sequence*: there is a map (d_3 for the experts) $\operatorname{Hom}_R(H_2(\Delta), R) \longrightarrow \mathbb{K}$.

Chapter 9
Derived Functors and Spectral Sequences

This chapter gives a short, intense introduction to derived functors and spectral sequences, two powerful but sometimes intimidating topics. It is the most technical chapter of the book. It is aimed at the reader who is a non-specialist, so contains basic definitions and theorems (sometimes without proof) as well as illustrative examples. The efficiency of the standard algorithm of Sect. 7.1.3 can be improved using insight from spectral sequences: see Sect. VII.4 of [67], and also [104], [105], [120]. We give an application of these tools to multiparameter persistent homology. Exercise 2.2.18 showed that tensoring a short exact sequence of R-modules

$$0 \longrightarrow A_1 \xrightarrow{f} A_2 \longrightarrow A_3 \longrightarrow 0$$

with an R-module M results in a sequence

$$A_1 \otimes_R M \xrightarrow{\overline{f}} A_2 \otimes_R M \longrightarrow A_3 \otimes_R M \longrightarrow 0$$

which is only exact on the right. What is the kernel of the induced map $\overline{f}$? Derived functors provide the answer; to set up the machinery we first need to define projective and injective objects, and resolutions. This chapter covers

- Injective and Projective Objects.
- Derived Functors.
- Spectral Sequences.
- Pas de deux: Spectral Sequences and Derived Functors.

Comprehensive treatments of the material here (and proofs) may be found in Eisenbud [70], Hartshorne [90], and Weibel [149]. Throughout this chapter, the ring R is a commutative Noetherian ring with unit over a field $\mathbb{K}$, and the sheaf $\mathcal{O}_X$ is the sheaf of regular functions on a variety X.

© The Author(s), under exclusive license to Springer Nature Switzerland AG 2022
H. Schenck, *Algebraic Foundations for Applied Topology and Data Analysis*,
Mathematics of Data 1, https://doi.org/10.1007/978-3-031-06664-1_9

9.1   Injective and Projective Objects, Resolutions

9.1.1   Projective and Injective Objects

Definition 9.1.1 A module P is **projective** if it possesses a universal lifting property: For any R–modules G and H, given a homomorphism $P \xrightarrow{\alpha} H$ and surjection $G \xrightarrow{\beta} H$, there exists a homomorphism θ making the diagram below commute:

$$
\begin{array}{ccc}
 & P & \\
 {\scriptstyle\theta}\swarrow_{\beta} & \downarrow{\scriptstyle\alpha} & \\
G \xrightarrow{\quad} H & \xrightarrow{\quad} & 0
\end{array}
$$

Recall that a finitely-generated R-module M is *free* if M is isomorphic to a direct sum of copies of the R–module R. Free modules are projective.

Definition 9.1.2 A module I is **injective** if given a homomorphism $H \xrightarrow{\alpha} I$ and injection $H \xrightarrow{\beta} G$, there exists a homomorphism θ making the diagram below commute:

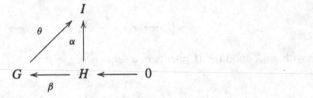

$$
\begin{array}{ccc}
 & I & \\
 {\scriptstyle\theta}\nearrow \uparrow{\scriptstyle\alpha} & & \\
G \xleftarrow[\beta]{\quad} H & \xleftarrow{\quad} & 0
\end{array}
$$

Projective and Injective modules will come to the forefront in the next section, which describes *derived functors*.

9.1.2   Resolutions

Given an R-module M, there exists a projective module surjecting onto M; for example, take a free module with a generator for each element of M. This yields an exact sequence:

$$P_0 \xrightarrow{d_0} M \longrightarrow 0.$$

The map d_0 has a kernel, so the process can be iterated, producing an exact sequence (possibly infinite) of free modules, terminating in M.

Definition 9.1.3 A *projective resolution* for an R–module M is an exact sequence of projective modules

$$\cdots P_2 \xrightarrow{d_2} P_1 \xrightarrow{d_1} P_0, \text{ with coker}(d_1) = M.$$

Notice there is no uniqueness property; for example we could set $P_2' = P_2 \oplus R$ and $P_1' = P_1 \oplus R$, and define a map $P_2' \to P_1'$ which is the identity on the R–summands, and the original map on the P_i summands. In the category of R–modules the construction above shows that projective resolutions always exist. Surprisingly this is not the case for sheaves of $\mathcal{O}_X$-modules. In fact ([90], Exercise III.6.2), for $X = \mathbb{P}^1$ there is no projective object surjecting onto $\mathcal{O}_X$.

Definition 9.1.4 An *injective resolution* for an R–module M is an exact sequence of injective modules

$$I_0 \xrightarrow{d_0} I_1 \xrightarrow{d_1} I_2, \cdots \text{ with ker}(d_0) = M.$$

It is not obvious that injective resolutions exist, but it can be shown (e.g. [70], [90], [149]) that in both the category of R–modules and in the category of sheaves of $\mathcal{O}_X$–modules, every object does have an injective resolution.

Exercise 9.1.5 Prove that in the category of $\mathbb{Z}$-modules (Abelian groups) that every object has an injective resolution. An injective $\mathbb{Z}$-module is called *divisible*; if you get stuck, see Sect. IV.3 of [93]. ◇

Exercise 9.1.6 Prove that in a category in which every object includes in an injective object, injective resolutions of complexes always exist. ◇

9.2 Derived Functors

In this section we describe the construction of derived functors, focusing on Ext^i (in the category of R-modules) and H^i (in the category of sheaves of $\mathcal{O}_X$–modules). For brevity we call these two categories "our categories". Working in our categories keeps things concrete and lets us avoid introducing too much terminology, while highlighting the most salient features of the constructions, most of which apply in much more general contexts. For proofs and a detailed discussion, see [70] or [149]. We quickly review the definitions from Chap. 7:

9.2.1 Categories and Functors

Recall that a category is a class of objects, along with morphisms between the objects, satisfying certain properties: composition of morphisms is associative, and identity morphisms exist.

Definition 9.2.1 Suppose $\mathscr{B}$ and $\mathscr{C}$ are categories. A *functor* F is a function from $\mathscr{B}$ to $\mathscr{C}$, taking objects to objects and morphisms to morphisms, preserving identity morphisms and compositions. If

$$B_1 \xrightarrow{b_1} B_2 \xrightarrow{b_2} B_3$$

is a sequence of objects and morphisms in $\mathscr{B}$, then

- F is *covariant* if applying F yields a sequence of objects and morphisms in $\mathscr{C}$ of the form:

$$F(B_1) \xrightarrow{F(b_1)} F(B_2) \xrightarrow{F(b_2)} F(B_3).$$

- F is *contravariant* if applying F yields a sequence of objects and morphisms in $\mathscr{C}$ of the form:

$$F(B_3) \xrightarrow{F(b_2)} F(B_2) \xrightarrow{F(b_1)} F(B_1).$$

A functor is *additive* if it preserves addition of homomorphisms; this property will be necessary in the construction of derived functors.

Example 9.2.2 The global sections functor of Chap. 3 is covariant: given a sequence of $\mathcal{O}_X$–modules

$$\mathscr{M}_1 \xrightarrow{f} \mathscr{M}_2 \xrightarrow{g} \mathscr{M}_3,$$

taking global sections yields a sequence

$$\Gamma(\mathscr{M}_1) \to \Gamma(\mathscr{M}_2) \to \Gamma(\mathscr{M}_3).$$

$\diamond$

Definition 9.2.3 Let F be a functor from $\mathscr{B}$ to $\mathscr{C}$, with $\mathscr{B}$ and $\mathscr{C}$ categories of modules over a ring. Let

$$0 \longrightarrow B_1 \xrightarrow{b_1} B_2 \xrightarrow{b_2} B_3 \longrightarrow 0$$

be a short exact sequence. F is *left–exact* if either

(a) F is covariant, and the sequence

$$0 \longrightarrow F(B_1) \xrightarrow{F(b_1)} F(B_2) \xrightarrow{F(b_2)} F(B_3) \quad \text{is exact, or}$$

(b) F is contravariant, and the sequence

$$0 \longrightarrow F(B_3) \xrightarrow{F(b_2)} F(B_2) \xrightarrow{F(b_1)} F(B_1) \quad \text{is exact.}$$

A similar definition applies for right exactness; a functor F is said to be exact if it is both left and right exact, which means that F preserves exact sequences.

9.2.2 Constructing Derived Functors

The construction of derived functors is motivated by the following question: if F is a left exact, contravariant functor and

$$0 \longrightarrow B_1 \xrightarrow{b_1} B_2 \xrightarrow{b_2} B_3 \longrightarrow 0$$

is a short exact sequence, then what is the cokernel of $F(b_1)$? This object will appear on the *right* of $F(B_1)$ in (b) above. A left exact functor has *right derived functors*. Similarly, a right exact functor has *left derived functors*.

Definition 9.2.4 Let $\mathscr{B}$ be the category of modules over a ring, and let F be a left exact, contravariant, additive functor from $\mathscr{B}$ to itself. If $M \in \mathscr{B}$, then there exists a projective resolution $P_\bullet$ for M.

$$\cdots \longrightarrow P_2 \xrightarrow{d_2} P_1 \xrightarrow{d_1} P_0$$

Applying F to $P_\bullet$ yields a complex:

$$0 \longrightarrow F(P_0) \xrightarrow{F(d_1)} F(P_1) \xrightarrow{F(d_2)} F(P_2) \longrightarrow \cdots .$$

The **right derived functors** $R^i F(M)$ are defined as

$$R^i F(M) = H^i(F(P_\bullet)).$$

Theorem 9.2.5 $R^i F(M)$ *is independent of the choice of projective resolution, and has the following properties:*

- $R^0 F(M) = F(M)$.
- *If M is projective then $R^i F(M) = 0$ if $i > 0$.*
- *A short exact sequence*

$$B_\bullet : 0 \to B_1 \xrightarrow{b_1} B_2 \xrightarrow{b_2} B_3 \to 0$$

gives rise to a long exact sequence

$$R^{j-1} F(B_3) \xrightarrow{R^{j-1}(F(b_2))} R^{j-1} F(B_2) \xrightarrow{R^{j-1}(F(b_1))} R^{j-1} F(B_1)$$

$$\delta_{j-1}$$

$$R^j F(B_3) \xrightarrow{R^j(F(b_2))} R^j F(B_2) \xrightarrow{R^j(F(b_1))} R^j F(B_1)$$

$$\delta_j$$

$$R^{j+1} F(B_3) \xrightarrow{R^{j+1}(F(b_2))} R^{j+1} F(B_2) \xrightarrow{R^{j+1}(F(b_1))} R^{j+1} F(B_1)$$

of derived functors, where the connecting maps are natural: given another short exact sequence $C_\bullet$ and map from $B_\bullet$ to $C_\bullet$, the obvious diagram involving the $R^i F$ commutes.

The proof is almost exactly the same as the proof of Theorem 4.3.6; see Proposition A3.17 of [70]. There are four possible combinations of variance and exactness; the type of resolution used to compute the derived functors of F is given below:

F	covariant	contravariant
left exact	injective	projective
right exact	projective	injective

Exercise 9.2.6 Prove that the derived functors do not depend on choice of resolution. The key to this is to construct a homotopy between resolutions, combined with the use of Theorem 4.3.6.   ◇

In the next sections, we study some common derived functors.

9.2.3 Ext

Let R be a ring, and suppose

$$B_\bullet : 0 \to B_1 \xrightarrow{b_1} B_2 \xrightarrow{b_2} B_3 \to 0$$

is a short exact sequence of R–modules, with N some fixed R–module. Applying $\text{Hom}_R(\bullet, N)$ to $B_\bullet$ yields an exact sequence:

$$0 \to \text{Hom}_R(B_3, N) \xrightarrow{c_1} \text{Hom}_R(B_2, N) \xrightarrow{c_2} \text{Hom}_R(B_1, N),$$

with $c_1(\phi) \mapsto \phi \circ b_2$ and $c_2(\theta) \mapsto \theta \circ b_1$; $\text{Hom}_R(\bullet, N)$ is left exact and contravariant.

Definition 9.2.7 $\text{Ext}^i_R(\bullet, N)$ is the ith right derived functor of $\text{Hom}_R(\bullet, N)$

Given R–modules M and N, to compute $\text{Ext}^i_R(M, N)$, we must find a projective resolution for M

$$\cdots \to P_2 \xrightarrow{d_2} P_1 \xrightarrow{d_1} P_0,$$

and compute the homology of the complex

$$0 \to \text{Hom}(P_0, N) \to \text{Hom}(P_1, N) \to \text{Hom}(P_2, N) \to \cdots$$

Example 9.2.8 Let $R = \mathbb{C}[x, y, z]$, $M = R/\langle xy, xz, yz \rangle$, and suppose $N \simeq R^1$. Applying $\text{Hom}_R(\bullet, R^1)$ to the projective (indeed, free) resolution of M

$$0 \longrightarrow R(-3)^2 \xrightarrow{\begin{bmatrix} -z & -z \\ y & 0 \\ 0 & x \end{bmatrix}} R(-2)^3 \xrightarrow{\begin{bmatrix} xy & xz & yz \end{bmatrix}} R$$

simply means dualizing the modules and transposing the differentials, so $\text{Ext}^i(R/I, R)$ is:

$$H_i \left[0 \longrightarrow R \xrightarrow{\begin{bmatrix} xy \\ xz \\ yz \end{bmatrix}} R(2)^3 \xrightarrow{\begin{bmatrix} -z & y & 0 \\ -z & 0 & x \end{bmatrix}} R(3)^2 \longrightarrow 0 \right]$$

Thus, $\text{Ext}^2(R/I, R)$ is the cokernel of the last map, and it is easy to check that $\text{Ext}^0(R/I, R) = \text{Ext}^1(R/I, R) = 0$. $\diamond$

For a fixed R–module M, applying $\mathrm{Hom}_R(M, \bullet)$ to $B_\bullet$ yields an exact sequence:

$$0 \longrightarrow \mathrm{Hom}_R(M, B_1) \xrightarrow{c_1} \mathrm{Hom}_R(M, B_2) \xrightarrow{c_2} \mathrm{Hom}_R(M, B_3) \ ,$$

with $c_1(\phi) \mapsto b_1 \circ \phi$ and $c_2(\theta) \mapsto b_2 \circ \theta$; $\mathrm{Hom}_R(\cdot, M)$ is left exact and covariant. Thus, to compute the derived functors of $\mathrm{Hom}_R(\cdot, M)$, on a module N, we must find an injective resolution of N:

$$I^0 \longrightarrow I^1 \longrightarrow I^2 \longrightarrow \cdots$$

then compute

$$H_i\left[0 \longrightarrow \mathrm{Hom}(I^0, M) \longrightarrow \mathrm{Hom}(I^1, M) \longrightarrow \mathrm{Hom}(I^2, M) \longrightarrow \cdots \right]$$

Using spectral sequences (next section), it is possible to show that $\mathrm{Ext}^i(M, N)$ can be regarded as the ith derived functor of *either* $\mathrm{Hom}_R(\bullet, N)$ or $\mathrm{Hom}_R(M, \bullet)$.

9.2.4   The Global Sections Functor

Let X be a variety, and suppose $\mathscr{B}$ is a coherent $\mathscr{O}_X$–module. The global sections functor Γ is left exact and covariant by Theorem 5.1.7. Hence, to compute $R^i\Gamma(\mathscr{B})$, we take an injective resolution of $\mathscr{B}$:

$$\mathscr{I}^0 \longrightarrow \mathscr{I}^1 \longrightarrow \mathscr{I}^2 \longrightarrow \cdots$$

then compute

$$H^i\left[0 \longrightarrow \Gamma(\mathscr{I}^0) \longrightarrow \Gamma(\mathscr{I}^1) \longrightarrow \Gamma(\mathscr{I}^2) \longrightarrow \cdots \right]$$

In Example 9.2.8 we wrote down an explicit free resolution and computed the Ext–modules. Unfortunately, the general construction for injective resolutions produces very complicated objects. For example, if $R = \mathbb{K}[x_1, \ldots, x_n]$, then the smallest injective R-module in which $\mathbb{K}$ can be included is infinitely generated.

It is not obvious that there is a relation between the Čech cohomology which appeared in Chap. 5 and the derived functors of Γ defined above. Later in this chapter, we'll see that there is a map

$$\check{H}^i(\mathscr{U}, \mathscr{F}) \longrightarrow H^i(X, \mathscr{F}),$$

and use spectral sequences to show that with certain conditions on $\mathcal{U}$ this is an isomorphism. The upshot is that many key facts about cohomology such as Theorem 5.1.7 follow naturally from the derived functor machinery.

9.2.5 Acyclic Objects

The last concept we need in order to work with derived functors is the notion of an acyclic object.

Definition 9.2.9 Let F be a left–exact, covariant functor. An object A is *acyclic* for F if $R^i F(A) = 0$ for all $i > 0$. An acyclic resolution of M is an exact sequence

$$A^0 \xrightarrow{d^0} A^1 \xrightarrow{d^1} A^2 \xrightarrow{d^2} \cdots$$

where the A^i are acyclic, and $M = \ker(d^0)$.

The reason acyclic objects are important is that a resolution of acyclic objects is good enough to compute higher derived functors; in other words we have an alternative to using resolutions by projective or injective objects.

Theorem 9.2.10 *Let $\mathcal{M}$ be a coherent $\mathcal{O}_X$–module, and*

$$\mathscr{A}^0 \longrightarrow \mathscr{A}^1 \longrightarrow \mathscr{A}^2 \longrightarrow \cdots$$

a Γ–acyclic resolution of $\mathcal{M}$. Then

$$R^i \Gamma(\mathcal{M}) = H^i \left[0 \longrightarrow \Gamma(\mathscr{A}^0) \longrightarrow \Gamma(\mathscr{A}^1) \longrightarrow \Gamma(\mathscr{A}^2) \longrightarrow \cdots \right]$$

Proof First, break the resolution into short exact sequences:

Since the $\mathscr{A}^i$ are acyclic for Γ, applying Γ to the short exact sequence

$$0 \to \mathscr{M} \to \mathscr{A}^0 \to \mathscr{M}^0 \to 0$$

yields an exact sequence

$$0 \to \Gamma(\mathscr{M}) \to \Gamma(\mathscr{A}^0) \to \Gamma(\mathscr{M}^0) \to R^1\Gamma(\mathscr{M}) \to 0$$

Now apply the snake lemma to the middle two columns of the (exact, commutative) diagram below.

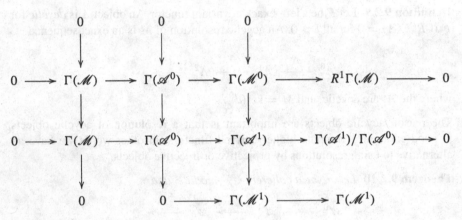

This yields a right exact sequence

$$0 \to R^1\Gamma(\mathscr{M}) \to \Gamma(\mathscr{A}^1)/\Gamma(\mathscr{A}^0) \to \Gamma(\mathscr{M}^1) \simeq \Gamma(\mathscr{A}^1)/\ker(d^1),$$

where $\Gamma(\mathscr{A}^1) \xrightarrow{d^1} \Gamma(\mathscr{A}^2)$. Hence,

$$R^1\Gamma(\mathscr{M}) = H^1 \Big[\ 0 \longrightarrow \Gamma(\mathscr{A}^0) \longrightarrow \Gamma(\mathscr{A}^1) \longrightarrow \Gamma(\mathscr{A}^2) \longrightarrow \cdots \ \Big]$$

In the next exercise you'll show that iterating this process yields the theorem. □

Exercise 9.2.11 Complete the proof of Theorem 9.2.10 by replacing the sequence

$$0 \to \mathscr{M} \to \mathscr{A}^0 \to \mathscr{M}^0 \to 0$$

with

$$0 \to \mathscr{M}^{i-1} \to \mathscr{A}^i \to \mathscr{M}^i \to 0$$

◇

9.3 Spectral Sequences

Spectral sequences are a fundamental tool in algebra and topology; when first encountered, they can seem quite confusing. In this brief overview, we describe a specific type of spectral sequence, state the main theorem, and illustrate the use of spectral sequences with several examples.

9.3.1 Total Complex of Double Complex

Definition 9.3.1 A *first quadrant double complex* is a commuting diagram, where each row and each column is a complex:

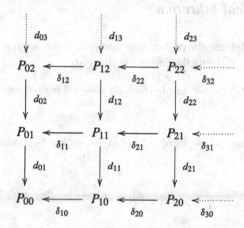

For each antidiagonal, define a module

$$P_m = \bigoplus_{i+j=m} P_{ij}.$$

We may define maps

$$P_m \xrightarrow{D_m} P_{m-1}$$

via

$$D_m(c_{ij}) = d_{ij}(c_{ij}) + (-1)^m \delta_{ij}(c_{ij}).$$

Thus, $D_{m-1}D_m(a) = dd(a) + \delta\delta(a) \pm (d\delta(a) - \delta d(a))$. The fact that each row and each column are complexes implies that $\delta\delta(a) = 0$ and $dd(a) = 0$. The commutativity of the diagram implies that $d\delta(a) = \delta d(a)$, and so $D^2 = 0$.

Definition 9.3.2 The *total complex* Tot(P) associated to a double complex P_{ij} is the complex $(\mathscr{P}_\bullet, D_\bullet)$ defined above.

Definition 9.3.3 A *filtration* of a module M is a chain of submodules

$$0 \subseteq M_n \subseteq M_{n-1} \subseteq \cdots \subseteq M_1 \subseteq M_0 = M$$

A filtration has an associated graded object $\mathrm{gr}(M) = \oplus M_i / M_{i+1}$. The main theorem concerning the spectral sequence of a double complex describes two different filtrations of the homology of the associated total complex. To describe these filtrations, we need to follow two different paths through the double complex.

9.3.2 The Vertical Filtration

For a double complex as above, we first compute homology with respect to the vertical differentials, yielding the following diagram:

$$\ker(d_{02})/\mathrm{im}(d_{03}) \xleftarrow{\;\delta_{12}\;} \ker(d_{12})/\mathrm{im}(d_{13}) \xleftarrow{\;\delta_{22}\;} \ker(d_{22})/\mathrm{im}(d_{23}) \xleftarrow{\cdots}$$

$$\ker(d_{01})/\mathrm{im}(d_{02}) \xleftarrow{\;\delta_{11}\;} \ker(d_{11})/\mathrm{im}(d_{12}) \xleftarrow{\;\delta_{21}\;} \ker(d_{21})/\mathrm{im}(d_{22}) \xleftarrow{\cdots}$$

$$P_{00}/\mathrm{im}(d_{01}) \xleftarrow{\;\delta_{10}\;} P_{10}/\mathrm{im}(d_{11}) \xleftarrow{\;\delta_{20}\;} P_{20}/\mathrm{im}(d_{21}) \xleftarrow{\cdots}$$

These objects are renamed as follows:

$$\mathrm{vert}E^1_{02} \xleftarrow{\;\delta_{12}\;} \mathrm{vert}E^1_{12} \xleftarrow{\;\delta_{22}\;} \mathrm{vert}E^1_{22} \xleftarrow[\;\delta_{32}\;]{}$$

$$\mathrm{vert}E^1_{01} \xleftarrow{\;\delta_{11}\;} \mathrm{vert}E^1_{11} \xleftarrow{\;\delta_{21}\;} \mathrm{vert}E^1_{21} \xleftarrow[\;\delta_{31}\;]{}$$

$$\mathrm{vert}E^1_{00} \xleftarrow{\;\delta_{10}\;} \mathrm{vert}E^1_{10} \xleftarrow{\;\delta_{20}\;} \mathrm{vert}E^1_{20} \xleftarrow[\;\delta_{30}\;]{}$$

The vertical arrows disappeared after computing homology with respect to d, and the horizontal arrows reflect the induced maps on homology from the original diagram. Now, compute the homology of the diagram above, with respect to the horizontal maps. For example, the object $\text{vert}E_{11}^2$ represents

$$\ker(\text{vert}E_{11}^1 \xrightarrow{\delta_{11}} \text{vert}E_{01}^1)/\text{im}(\text{vert}E_{21}^1 \xrightarrow{\delta_{21}} \text{vert}E_{11}^1)$$

The resulting modules may be displayed in a grid:

$$\text{vert}E_{02}^2 \qquad\qquad \text{vert}E_{12}^2 \qquad\qquad \text{vert}E_{22}^2$$

$$\text{vert}E_{01}^2 \qquad\qquad \text{vert}E_{11}^2 \qquad\qquad \text{vert}E_{21}^2$$

$$\text{vert}E_{00}^2 \qquad\qquad \text{vert}E_{10}^2 \qquad\qquad \text{vert}E_{20}^2$$

Although it appears at first that there are no maps between these objects, the crucial observation is that there is a map $d_{i,j}^2$ from E_{ij}^2 to $E_{i-2,j+1}^2$. This "knight's move" is constructed just like the connecting map δ appearing in the snake lemma. The diagram above (with differentials added) is thus:

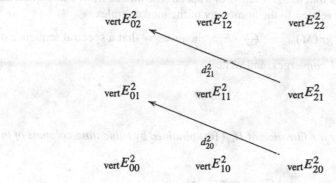

So we may compute homology with respect to this differential. The homology at position (i, j) is labeled, as one might expect, $\text{vert}E_{ij}^3$; it is now the case (but far from intuitive) that there is a differential $d_{i,j}^3$ taking $\text{vert}E_{ij}^3$ to $\text{vert}E_{i-3,j+2}^3$:

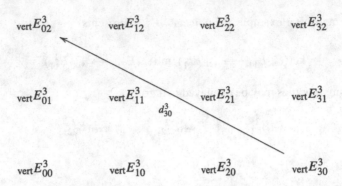

$$\text{vert}\,E_{02}^3 \qquad\qquad \text{vert}\,E_{12}^3 \qquad\qquad \text{vert}\,E_{22}^3 \qquad\qquad \text{vert}\,E_{32}^3$$

$$\text{vert}\,E_{01}^3 \qquad\qquad \text{vert}\,E_{11}^3 \qquad\qquad \text{vert}\,E_{21}^3 \qquad\qquad \text{vert}\,E_{31}^3$$

$$\text{vert}\,E_{00}^3 \qquad\qquad \text{vert}\,E_{10}^3 \qquad\qquad \text{vert}\,E_{20}^3 \qquad\qquad \text{vert}\,E_{30}^3$$

The process continues, with d_{ij}^r mapping $\text{vert}\,E_{ij}^r$ to $\text{vert}\,E_{i-r,j+r-1}^r$. One thing that is obvious is that since the double complex lies in the first quadrant, eventually the differentials in and out at position (i, j) must be zero, so that the module at position (i, j) stabilizes; it is written $\text{vert}\,E_{ij}^\infty$. For example, it is easy to see that $\text{vert}\,E_{10}^2 = \text{vert}\,E_{10}^\infty$, while $\text{vert}\,E_{20}^2 \neq \text{vert}\,E_{20}^\infty$ but $\text{vert}\,E_{20}^3 = \text{vert}\,E_{20}^\infty$.

9.3.3   Main Theorem

The main theorem is that the E^∞ terms of a spectral sequence from a first quadrant double complex are related to the homology of the total complex.

Definition 9.3.4 If $gr(M)_m \simeq \bigoplus\limits_{i+j=m} E_{ij}^\infty$, then we say that a spectral sequence of the filtered object M *converges*, and write

$$E^r \Rightarrow M$$

Theorem 9.3.5 *For the filtration of $H_m(\text{Tot})$ obtained by truncating columns of the double complex,*

$$\bigoplus\limits_{i+j=m} \text{vert}\,E_{ij}^\infty \Rightarrow H_m(\text{Tot}).$$

As with the long exact sequence of derived functors, the proof is not bad, but lengthy, so we refer to [149] or [70] for details. In Sect. 3.2, we first computed homology with respect to the vertical differential d. If instead we first compute homology with respect to the horizontal differential δ, then the higher differentials are:

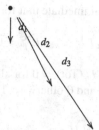

As before, for $r \gg 0$, the source and target are zero, so the homology at position (i, j) stabilizes. The resulting value is denoted $_{\mathrm{hor}} E_{ij}^{\infty}$, and we have:

Theorem 9.3.6 *For the filtration of $H_m(\mathrm{Tot})$ obtained by truncating rows of the double complex,*

$$\bigoplus_{i+j=m} {}_{\mathrm{hor}} E_{ij}^{\infty} \Rightarrow H_m(\mathrm{Tot}).$$

For a first quadrant double complex, the two theorems above tell us that

$$\bigoplus_{i+j=m} {}_{\mathrm{hor}} E_{ij}^{\infty} \Rightarrow H_m(\mathrm{Tot}) \quad \text{and} \quad \bigoplus_{i+j=m} {}_{\mathrm{vert}} E_{ij}^{\infty} \Rightarrow H_m(\mathrm{Tot}).$$

Because the filtrations for the horizontal and vertical spectral sequence are different, it is often the case that for one of the spectral sequences the E^{∞} terms stabilize very early, perhaps even vanishing. So the main idea is to play off the two different filtrations against each other. This is illustrated in the next example.

Example 9.3.7 We now give a spectral sequence proof of Theorem 4.3.6. Let $0 \to C_2 \to C_1 \to C_0 \to 0$ be a short exact sequence of complexes:

$$
\begin{array}{ccccccccc}
& & 0 & & 0 & & 0 & & \\
& & \downarrow & & \downarrow & & \downarrow & & \\
C_2:0 & \longleftarrow & C_{02} & \longleftarrow & C_{12} & \longleftarrow & C_{22} & \longleftarrow\!\cdots & \\
& & \downarrow & & \downarrow & & \downarrow & & \\
C_1:0 & \longleftarrow & C_{01} & \longleftarrow & C_{11} & \longleftarrow & C_{21} & \longleftarrow\!\cdots & \\
& & \downarrow & & \downarrow & & \downarrow & & \\
C_0:0 & \longleftarrow & C_{00} & \longleftarrow & C_{10} & \longleftarrow & C_{20} & \longleftarrow\!\cdots & \\
& & \downarrow & & \downarrow & & \downarrow & & \\
& & 0 & & 0 & & 0 & &
\end{array}
$$

Since the columns are exact, it is immediate that for all (i, j)

$$\text{vert} E_{ij}^1 = \text{vert} E_{ij}^\infty = 0$$

By Theorem 9.3.5, we conclude $H_m(\text{Tot}) = 0$ for all m. For the horizontal filtration $\text{hor} E_{ij}^1 = H_i(C_j)$ if $j \in \{0, 1, 2\}$, and 0 otherwise. For E^2 we have

$$\text{hor} E_{ij}^2 = \begin{cases} \ker(H_i(C_2) \to H_i(C_1)) & j = 2 \\ \ker(H_i(C_1) \to H_i(C_0))/\text{im}(H_i(C_2) \to H_i(C_1)) & j = 1 \\ \text{coker}(H_i(C_1) \to H_i(C_0)) & j = 0. \end{cases}$$

The d_2 differential is zero for the middle row, and maps $\text{hor}E_{i,2}^2 \to \text{hor}E_{i+1,0}^2$:

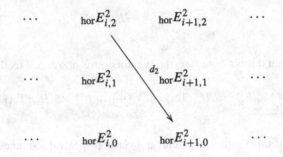

So $\text{hor}E_{i,1}^2 = \text{hor}E_{i,1}^\infty$, while

$$\text{hor}E_{i,2}^3 = \text{hor}E_{i,2}^\infty = \ker(\text{hor}E_{i,2}^2 \to \text{hor}E_{i+1,0}^2)$$

and

$$\text{hor}E_{i,0}^3 = \text{hor}E_{i,0}^\infty = \text{coker}(\text{hor}E_{i,2}^2 \to \text{hor}E_{i+1,0}^2)$$

By Theorem 9.3.6,

$$H_m(\text{Tot}) = \bigoplus_{i+j=m} \text{hor} E_{ij}^\infty$$

From the vertical spectral sequence, $H_m(\text{Tot}) = 0$, so all the terms $\text{hor} E_{ij}^\infty$ must vanish. Working backwards, we see this means

$$0 = \text{hor}E_{i,1}^2 = \ker(H_i(C_1) \to H_i(C_0))/\text{im}(H_i(C_2) \to H_i(C_1)),$$

hence $H_i(C_2) \to H_i(C_1) \to H_i(C_0)$ is exact, and

$$\ker(H_i(C_2) \to H_i(C_1)) \simeq \mathrm{coker}(H_{i+1}(C_1) \to H_{i+1}(C_0))$$

which yields the long exact sequence in homology.

Exercise 9.3.8 Tensor product is right exact and covariant. Fix an R-module N. Prove that the ith left derived functor of $\bullet \otimes_R N$ is isomorphic to the ith left derived functor of $N \otimes_R \bullet$ as follows: Let

$$\cdots \to P_2 \xrightarrow{p_2} P_1 \xrightarrow{p_1} P_0$$

be a projective resolution for M and

$$\cdots \to Q_2 \xrightarrow{q_2} Q_1 \xrightarrow{q_1} Q_0$$

be a projective resolution for N. Form the double complex

$$
\begin{array}{ccccccc}
\Big\downarrow{\scriptstyle d_{03}} & & \Big\downarrow{\scriptstyle d_{13}} & & \Big\downarrow{\scriptstyle d_{23}} & & \\
P_0 \otimes Q_2 & \xleftarrow{\ \delta_{12}\ } & P_1 \otimes Q_2 & \xleftarrow{\ \delta_{22}\ } & P_2 \otimes Q_2 & \xleftarrow{\ \ \ } & \\
\Big\downarrow{\scriptstyle d_{02}} & & \Big\downarrow{\scriptstyle d_{12}} & & \Big\downarrow{\scriptstyle d_{22}} & & \\
P_0 \otimes Q_1 & \xleftarrow{\ \delta_{11}\ } & P_1 \otimes Q_1 & \xleftarrow{\ \delta_{21}\ } & P_2 \otimes Q_1 & \xleftarrow{\ \ \ } & \\
\Big\downarrow{\scriptstyle d_{01}} & & \Big\downarrow{\scriptstyle d_{11}} & & \Big\downarrow{\scriptstyle d_{21}} & & \\
P_0 \otimes Q_0 & \xleftarrow{\ \delta_{10}\ } & P_1 \otimes Q_0 & \xleftarrow{\ \delta_{20}\ } & P_2 \otimes Q_0 & \xleftarrow{\ \ \ } &
\end{array}
$$

with differentials $P_i \otimes Q_j \xrightarrow{\delta_{ij}} P_{i-1} \otimes Q_j$ defined by $a \otimes b \mapsto p_i(a) \otimes b$, and $P_i \otimes Q_j \xrightarrow{d_{ij}} P_i \otimes Q_{j-1}$ defined by $a \otimes b \mapsto a \otimes q_j(b)$.

(a) Show that for the vertical filtration, the E^1 terms are

$$
{}_{vert}E^1_{ij} = \begin{cases} P_i \otimes N & j = 0 \\ 0 & j \neq 0. \end{cases}
$$

Now explain why $_{vert}E^2 = {}_{vert}E^\infty$, and these terms are:

$$_{vert}E^2_{ij} = \begin{cases} H_i(P_\bullet \otimes N) = Tor_i(M, N) & j = 0 \\ 0 & j \neq 0. \end{cases}$$

(b) Show that for the horizontal filtration

$$_{hor}E^2_{ij} = \begin{cases} H_j(M \otimes Q_\bullet) = Tor_j(N, M) & i = 0 \\ 0 & i \neq 0. \end{cases}$$

(c) Put everything together to conclude that

$$\begin{aligned} \mathrm{Tor}_m(M, N) &\simeq \bigoplus_{i+j=m} \text{vert } E^\infty_{ij} \\ &\simeq gr(H_m(\mathrm{Tot})) \\ &\simeq \bigoplus_{i+j=m} \text{hor } E^\infty_{ij} \\ &\simeq \mathrm{Tor}_m(N, M) \end{aligned}$$

This fact is sometimes expressed by saying that Tor is a *balanced* functor. ◇

Exercise 9.3.9 Prove that $\mathrm{Ext}^i(M, N)$ can be regarded as the ith derived functor of *either* $\mathrm{Hom}_R(\bullet, N)$ or $\mathrm{Hom}_R(M, \bullet)$. The method is quite similar to the proof above, except for this one, you'll need both projective and injective resolutions. ◇

9.4 Pas de Deux: Spectral Sequences and Derived Functors

In this last section, we'll see how useful the machinery of spectral sequences is in yielding theorems about derived functors. To do this, we first need to define resolutions of complexes. Note that sometimes our differentials on the double complex go "up and right" instead of "down and left", so the higher differentials will change accordingly.

9.4.1 Resolution of a Complex

Suppose

$$A : 0 \longrightarrow \mathscr{A}^0 \longrightarrow \mathscr{A}^1 \longrightarrow \mathscr{A}^2 \longrightarrow \cdots$$

is a complex, either of R–modules or of sheaves of $\mathcal{O}_X$–modules. An *injective resolution* of A is a double complex as below:

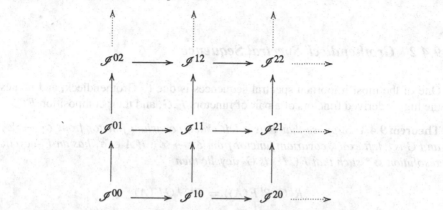

Let d^{jk} denote the horizontal differential at position (j, k). The double complex is a resolution if it satisfies the following properties

(a) The complex is exact.
(b) Each column $\mathcal{I}^{i*}$ is an injective resolution of $\mathcal{A}^i$.
(c) $\ker(d^{jk})$ is an injective summand of $\mathcal{I}^{jk}$.

Condition (c) implies that $\operatorname{im}(d^{j,k})$ is injective, yielding a "Hodge decomposition":

$$0 \longrightarrow \operatorname{im}(d^{j-1,k}) \longrightarrow \ker(d^{j,k}) \longrightarrow H_d^{j,k} \longrightarrow 0$$

$$\operatorname{im}(d^{j-1,k})$$

It follows that we may decompose the sequence

$$\mathcal{I}^{j-1,k} \xrightarrow{\ d^{j-1,k}\ } \mathcal{I}^{j,k} \xrightarrow{\ d^{j,k}\ } \mathcal{I}^{j+1,k}$$

as:

$$\cdots \rightarrow \operatorname{im}(d^{j-2,k}) \xrightarrow{\ 0\ } \operatorname{im}(d^{j,k}) \xrightarrow{\ 1\ } \operatorname{im}(d^{j,k}) \rightarrow \cdots$$

$$\oplus \qquad\qquad \oplus \qquad\qquad \oplus$$

$$\cdots \rightarrow H^{j-1,k} \xrightarrow{\ 0\ } H^{j,k} \xrightarrow{\ 0\ } H^{j+1,k} \rightarrow \cdots$$

$$\oplus \qquad\qquad \oplus \qquad\qquad \oplus$$

$$\cdots \rightarrow \operatorname{im}(d^{j-1,k}) \xrightarrow{\ 1\ } \operatorname{im}(d^{j-1,k}) \xrightarrow{\ 0\ } \operatorname{im}(d^{j+1,k}) \rightarrow \cdots$$

An inductive argument (Exercise 3.1) shows that in a category in which any object includes in an injective object, injective resolutions of complexes always exist.

9.4.2   Grothendieck Spectral Sequence

One of the most important spectral sequences is due to Grothendieck, and relates the higher derived functors of a pair of functors F, G, and their composition FG.

Theorem 9.4.1 *Suppose that F is a left exact, covariant functor from $\mathscr{C}_1 \to \mathscr{C}_2$ and G is a left exact, covariant functor from $\mathscr{C}_2 \to \mathscr{C}_3$. If $A \in \mathscr{C}_1$ has an F–acyclic resolution $\mathscr{A}^{\bullet}$ such that $F(\mathscr{A}^i)$ is G–acyclic then*

$$R^j G(R^i F(A)) \Rightarrow R^{i+j} GF(A)$$

Proof Take an injective resolution $\mathscr{I}^{\bullet,\bullet}$ for the complex

$$0 \longrightarrow F(\mathscr{A}^0) \longrightarrow F(\mathscr{A}^1) \longrightarrow F(\mathscr{A}^2) \longrightarrow \cdots$$

Apply G to $\mathscr{I}^{\bullet,\bullet}$. It follows from the construction above that a row of the double complex $G(\mathscr{I}^{\bullet,\bullet})$ has the form:

$$\cdots\dashrightarrow G(\mathrm{im}(d^{j-2,k})) \xrightarrow{G(0)} G(\mathrm{im}(d^{j,k})) \xrightarrow{G(1)} G(\mathrm{im}(d^{j,k})) \dashrightarrow$$

$$\oplus \qquad\qquad \oplus \qquad\qquad \oplus$$

$$\cdots\dashrightarrow G(H^{j-1,k}) \xrightarrow{G(0)} G(H^{j,k}) \xrightarrow{G(0)} G(H^{j+1,k}) \dashrightarrow$$

$$\oplus \qquad\qquad \oplus \qquad\qquad \oplus$$

$$\cdots\dashrightarrow G(\mathrm{im}(d^{j-1,k})) \xrightarrow{G(1)} G(\mathrm{im}(d^{j-1,k})) \xrightarrow{G(0)} G(\mathrm{im}(d^{j+1,k})) \dashrightarrow$$

Hence,

$$_{hor}E^1_{ij} = G(H^{i,j})$$

By construction, $H^{i,j}$ is the jth object in an injective resolution for the ith cohomology of $F(\mathscr{A}^{\bullet})$. Since $\mathscr{A}^{\bullet}$ was an F–acyclic resolution for A, the ith cohomology is exactly $R^i F(A)$, so that

$$_{hor}E^2_{ij} = H^j\left[0 \to G(H^{i,0}) \to G(H^{i,1}) \to G(H^{i,2}) \to \cdots\right] = R^j G(R^i F(A))$$

Next, we turn to the vertical filtration. We have the double complex

$$
\begin{array}{ccccc}
\vdots & & \vdots & & \vdots \\
\uparrow & & \uparrow & & \uparrow \\
G(\mathscr{I}^{02}) & \longrightarrow & G(\mathscr{I}^{12}) & \longrightarrow & G(\mathscr{I}^{22}) \dashrightarrow \\
\uparrow & & \uparrow & & \uparrow \\
G(\mathscr{I}^{01}) & \longrightarrow & G(\mathscr{I}^{11}) & \longrightarrow & G(\mathscr{I}^{21}) \dashrightarrow \\
\uparrow & & \uparrow & & \uparrow \\
G(\mathscr{I}^{00}) & \longrightarrow & G(\mathscr{I}^{10}) & \longrightarrow & G(\mathscr{I}^{20}) \dashrightarrow
\end{array}
$$

Since $\mathscr{I}^{ij}$ is an injective resolution of $F(\mathscr{A}^i)$,

$$
R^j G(F(\mathscr{A}^i)) = H^j \left[0 \to G(\mathscr{I}^{i0}) \to G(\mathscr{I}^{i1}) \to G(\mathscr{I}^{i2}) \to \cdots \right].
$$

Now, the assumption that the $F(\mathscr{A}^i)$ are G–acyclic forces $R^j G(F(\mathscr{A}^i))$ to vanish, for all $j > 0$! Hence, the cohomology of a *column* of the double complex above vanishes, except at position zero. In short

$$
\mathrm{vert} E^1_{ij} = \begin{cases} GF(\mathscr{A}^i) & j = 0 \\ 0 & j \neq 0. \end{cases}
$$

Thus,

$$
\mathrm{vert} E^\infty_{ij} = \mathrm{vert} E^2_{ij} = \begin{cases} R^{i+j} GF(A) & j = 0 \\ 0 & j \neq 0. \end{cases}
$$

Applying Theorems 9.3.5 and 9.3.6 concludes the proof. $\qquad\qquad\square$

Exercise 9.4.2 Let $Y \xrightarrow{f} X$ be a continuous map between topological spaces, with $\mathscr{F}$ a sheaf on Y. The pushforward is defined via:

$$
f_* \mathscr{F}(V) = \mathscr{F}(f^{-1}(V))
$$

(a) Show that pushforward f_* is left exact and covariant, so associated to $\mathscr{F}$ are objects $R^j f_*(\mathscr{F})$.

(b) Use Theorem 9.4.1 to obtain the *Leray spectral sequence*:

$$H^i(R^j f_*(\mathscr{F})) \Rightarrow H^{i+j}(\mathscr{F})$$

What can you say when Y is the fiber of a vector bundle X?   ◇

9.4.3 Comparing Cohomology Theories

Our second application of spectral sequences will be to relate the higher derived functors of Γ to the Čech cohomology. If I_p denotes a $p+1$-tuple $\{i_0 < i_1 < \cdots < i_p\}$ and $U_{I_p} = U_{i_0} \cap \cdots \cap U_{i_p}$, then applying the pushforward construction from Exercise 9.4.2 to the inclusion $U_{I_p} \overset{i}{\hookrightarrow} X$ gives a sheaf theoretic version of the Čech complex.

$$\mathscr{C}^p(\mathscr{U}, \mathscr{F}) = \prod_{I_p} i_*(\mathscr{F}|_{U_{I_p}}).$$

Exercise 9.4.3 Show that $\mathscr{C}^\bullet$ is a resolution of $\mathscr{F}$, as follows. By working at the level of stalks, show that there is a morphism of complexes

$$\mathscr{C}^i(\mathscr{U}, \mathscr{F})_p \overset{k}{\longrightarrow} \mathscr{C}^{i-1}(\mathscr{U}, \mathscr{F})_p$$

such that $(d_{i-1}k + kd_i)$ is the identity. Conclude by applying Theorem 4.3.8. Finally, show that if $\mathscr{F}$ is injective, then so are the sheaves $\mathscr{C}^i(\mathscr{U}, \mathscr{F})$. If you get stuck, see [90], III.4.   ◇

Lemma 9.4.4 *For an open cover $\mathscr{U}$, there is a map $\check{H}^i(\mathscr{U}, \mathscr{F}) \longrightarrow H^i(X, \mathscr{F})$.*

Proof Take an injective resolution $\mathscr{I}^\bullet$ for $\mathscr{F}$. By injectivity, we get

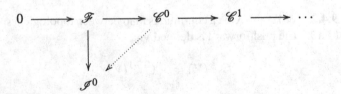

Iterating the construction gives a map of complexes $\mathscr{C}^\bullet \to \mathscr{I}^\bullet$, and it follows from Lemma 4.3.4 that there is a corresponding map on cohomology.   □

Theorem 9.4.5 *Let $\mathscr{U}$ be a* Leray cover: *an open cover such that for any I_p,*

$$H^i(U_{I_p}, \mathscr{F}) = 0, \text{ for all } i \geq 1.$$

Then

$$\check{H}^i(\mathscr{U}, \mathscr{F}) = H^i(X, \mathscr{F}).$$

Proof Take an injective resolution $\mathscr{I}^\bullet$ for $\mathscr{F}$. Since $H^i(U_{I_p}, \mathscr{F}) = 0, i > 0$

$$0 \longrightarrow \mathscr{F}(U_{I_p}) \longrightarrow \mathscr{I}^0(U_{I_p}) \longrightarrow \mathscr{I}^1(U_{I_p}) \longrightarrow \cdots$$

is exact. Then as in the construction of the sheaf-theoretic Čech complex, we obtain a Čech complex built out of the direct product of these, which is by construction a resolution (depicted below) of the Čech complex for $\mathscr{F}$. The bottom row is included for clarity, it is *not* part of the complex.

$$
\begin{array}{ccccccc}
\uparrow & & \uparrow & & \uparrow & & \\
\mathscr{C}^0(\mathscr{U}, \mathscr{I}^2) & \rightarrow & \mathscr{C}^1(\mathscr{U}, \mathscr{I}^2) & \rightarrow & \mathscr{C}^2(\mathscr{U}, \mathscr{I}^2) & \dashrightarrow & \\
\uparrow & & \uparrow & & \uparrow & & \\
\mathscr{C}^0(\mathscr{U}, \mathscr{I}^1) & \rightarrow & \mathscr{C}^1(\mathscr{U}, \mathscr{I}^1) & \rightarrow & \mathscr{C}^2(\mathscr{U}, \mathscr{I}^1) & \dashrightarrow & \\
\uparrow & & \uparrow & & \uparrow & & \\
\mathscr{C}^0(\mathscr{U}, \mathscr{I}^0) & \rightarrow & \mathscr{C}^1(\mathscr{U}, \mathscr{I}^0) & \rightarrow & \mathscr{C}^2(\mathscr{U}, \mathscr{I}^0) & \dashrightarrow & \\
\uparrow & & \uparrow & & \uparrow & & \\
\mathscr{C}^0(\mathscr{U}, \mathscr{F}) & \longrightarrow & \mathscr{C}^1(\mathscr{U}, \mathscr{F}) & \longrightarrow & \mathscr{C}^2(\mathscr{U}, \mathscr{F}) & \dashrightarrow & \\
\end{array}
$$

Applying Γ, since $H^i(U_{I_p}, \mathscr{F}) = 0$ for $i > 0$,

$$_{vert}E^1_{ij} = \begin{cases} \Gamma(\mathscr{C}^i(\mathscr{U}, \mathscr{F})) & j = 0 \\ 0 & j \neq 0. \end{cases}$$

Thus, $E^2 = E^\infty$, and since $\Gamma(\mathscr{C}^i(\mathscr{U}, \mathscr{F})) = C^i(\mathscr{U}, \mathscr{F})$

$$_{vert}E^2_{ij} = \begin{cases} \check{H}^i(\mathscr{U}, \mathscr{F}) & j = 0 \\ 0 & j \neq 0. \end{cases}$$

For the horizontal filtration, since the $\mathscr{C}^i(\mathscr{U}, \mathscr{I}^j)$ are injective,

$$_{hor}E^1_{ij} = \begin{cases} \Gamma(\mathscr{I}^j) & i = 0 \\ 0 & i \neq 0. \end{cases}$$

and thus

$$_{hor}E^2_{ij} = \begin{cases} H^j(\Gamma(\mathscr{I}^\bullet)) & i = 0 \\ 0 & i \neq 0. \end{cases}$$

This is the usual derived functor cohomology, and applying Theorems 9.3.5 and 9.3.6 concludes the proof. $\square$

9.4.4 Cartan–Eilenberg Resolution

A *Cartan–Eilenberg* resolution of a complex is a resolution constructed in a fashion similar to the injective resolution of a complex appearing in Sect. 9.4.1. For a complex $C_\bullet$, take free resolutions $F_{i+1,\bullet}$ for $\mathrm{im}(\partial_{i+1})$ and $G_{i,\bullet}$ for $H_i(C)$. The short exact sequence

$$0 \longrightarrow \mathrm{im}(\partial_{i+1}) \longrightarrow \ker(\partial_i) \longrightarrow H_i(C) \longrightarrow 0$$

combined with the resolutions above yields a resolution $G_{i,\bullet} \oplus F_{i+1,\bullet}$ for $\ker(\partial_i)$. Now use the resolutions for $\ker(\partial_i)$ and $\mathrm{im}(\partial_i)$ and the short exact sequence

$$0 \longrightarrow \ker(\partial_i) \longrightarrow C_i \longrightarrow \mathrm{im}(\partial_i) \longrightarrow 0$$

to obtain a resolution for C_i. The resulting double complex has terms

$$F_{i+1,j} \oplus G_{i,j} \oplus F_{i,j} \xrightarrow{d_{i,j}} F_{i,j} \oplus G_{i-1,j} \oplus F_{i-1,j}$$

with $d_{i,j}$ the identity on $F_{i,j}$ and zero on the other summands.

Example 9.4.6 Our final spectral sequence example comes from the three-parameter filtration of S^2 appearing in Example 8.4.3

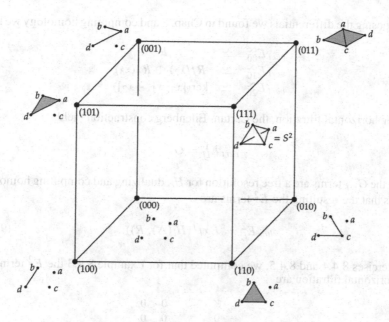

with edges in degrees

$$ab = 001 \quad ac = 011 \quad ad = 001 \quad bc = 010 \quad bd = 100 \quad cd = 010$$

and triangles in degrees

$$abc = 011 \quad abd = 101 \quad acd = 011 \quad bcd = 110.$$

Take a Cartan–Eilenberg resolution of the multiparameter persistent homology complex appearing in Definition 8.1.3. Now apply $Hom_R(\cdot, R)$, and compute the spectral sequences for the resulting double complex. The vertical filtration degenerates immediately, because

$$_{vert}E^1_{ij} = Ext^j(C_i, R) = \begin{cases} Hom_R(C_i, R) & j = 0 \\ 0 & j \neq 0, \end{cases}$$

which follows since the C_i are free modules, so $Ext^j(C_i, R)$ vanishes if $j \neq 0$. Let $C^*_\bullet$ be the complex of modules $C^*_i = Hom_R(C_i, R)$, we have

$$_{vert}E^2_{ij} = {}_{vert}E^\infty_{ij} = \begin{cases} H_i(C^*_\bullet) & j = 0 \\ 0 & j \neq 0, \end{cases} \tag{9.4.1}$$

Transposing the differentials we found in Chap. 8 and computing homology we have

$$
\begin{aligned}
_{vert}E_{00}^\infty &\simeq & R^1 \\
_{vert}E_{10}^\infty &\simeq & R/\langle x_2 \rangle \oplus R/\langle x_3 \rangle \\
_{vert}E_{20}^\infty &\simeq & \ker([x_3, x_1, -x_2])
\end{aligned}
$$

For the horizontal filtration, the Cartan–Eilenberg construction yields

$$
_{vert}E_{ij}^1 = G_{i,j}^*.
$$

Since the $G_{i,j}$ terms are a free resolution for H_i, dualizing and computing homology means that the resulting the E^2 terms are

$$
_{hor}E_{ij}^2 = Ext^j(H_i(\Delta), R)). \tag{9.4.2}
$$

In Exercises 8.4.4 and 8.4.5, we computed that for Example 8.4.3 the E^2 terms for the horizontal filtration are

$$
\begin{array}{ccc}
\mathbb{K} & 0 & 0 \\
0 & 0 & 0 \\
Ext^1(H_0(\Delta), R) & 0 & 0 \\
R^1 & 0 & R^1
\end{array}
$$

The bottom row follows from $\mathrm{Hom}_R(H_0(\Delta), R) \simeq R^1 \simeq \mathrm{Hom}_R(H_2(\Delta), R)$. There is a potentially nonzero d_3 differential

$$
\mathrm{Hom}_R(H_2(\Delta), R) \simeq R^1 \xrightarrow{d_3} Ext_R^3(H_0(\Delta), R) \simeq \mathbb{K}.
$$

To analyze this map, we use that the horizontal and vertical filtrations both yield associated graded objects for the direct sum of the diagonals. And indeed,

$$
\begin{aligned}
_{vert}E_{00}^\infty &\simeq & R^1 & \simeq & _{hor}E_{00}^\infty \\
_{vert}E_{10}^\infty &\simeq & R/\langle x_2 \rangle \oplus R/\langle x_3 \rangle = Ext^1(H_0(\Delta), R) & \simeq & _{hor}E_{01}^\infty
\end{aligned}
$$

Now we come to the interesting comparison. For the vertical filtration

$$
\bigoplus_{i+j=2} {}_{vert}E_{ij}^\infty \simeq \ker([x_3, x_1, -x_2])
$$

But this must be the associated graded of a filtration for $H_2(Tot)$. On the other hand,

$$
\bigoplus_{i+j=2} {}_{hor}E_{ij}^\infty \simeq \ker(d_3)
$$

So this means that we have a nontrivial d_3 differential mapping

$$R^1 \to \mathbb{K} = R^1/\langle x_1, x_2, x_3 \rangle,$$

yielding

$$\ker(d_3) = \ker([x_3, x_1, -x_2])$$

By Theorem 8.4.1 the modules $\text{Ext}^j(H_i(\Delta), R)$ encode the codimension j support locus of $H_i(\Delta)$. This means they provide a natural way to stratify MPH modules.

The Hilbert Syzygy Theorem shows that the resolution for a $\mathbb{K}[x_1, \ldots, x_n]$-module has length at most n, so

$$\text{Ext}^j(H_i(\Delta), R) = 0$$

when $j \geq n + 1$. Combining this with Eqs. 9.4.1 and 9.4.2 proves

Theorem 9.4.7 *For a Cartan–Eilenberg resolution of the MPH complex $C_\bullet$ of Definition 8.1.3, we have*

$$\bigoplus_{i+j=k} {}_{hor}E_{ij}^\infty \Longrightarrow H_k(C_\bullet^*),$$

and

$$_{hor}E_{ij}^m = {}_{hor}E_{ij}^\infty \text{ for all } m \geq n + 1.$$

Exercise 9.4.8 Carry out this analysis for other examples. Notice for a filtration with r−parameters, $\text{Ext}^j(H_i(\Delta), R)$ vanishes for $j > r$, which means that for the horizontal filtration the differentials d_j vanish for $j > r$. ◇

Appendix A
Examples of Software Packages

We include some examples of various software packages that perform the computations discussed in this book. Software is dynamic, so at some point the syntax for the examples below will be outdated. Nevertheless, the general structure of the computations is illustrative, and worth including.

A.1 Covariance and Spread of Data via R

We begin with a "barehanded" approach: in Exercise 1.3.10, we considered the dataset

$$X = \{(1, 1), (2, 2), (2, 3), (3, 2), (3, 3), (4, 4)\}$$

We found the covariance matrix, and computed the eigenvalues and eigenvectors. To visualize the ellipse that best fits the data, we use a standard statistics package: R, which is public domain, available at

```
https://www.r-project.org

# next commands load the visualization package
library(ggplot2)
library(ggpubr)

# next commands input data; if large data set load from file
x = c(1, 2, 2, 3, 3, 4)
y = c(1, 2, 3, 2, 3, 4)

datum = cbind(x, y)
datum = as.data.frame(datum)

# next commands get the covariance matrix
```

```
# note: without the second line, the result will not display

covariance = cov(datum)
covariance

# the result is displayed as

    x   y
x 1.1 0.9
y 0.9 1.1

# next commands get the eigenvalues and eigenvectors
eigenvalues = eigen(covariance)
eigenvalues

# the result is displayed as

eigen() decomposition
$values
[1] 2.0 0.2

$vectors
          [,1]        [,2]
[1,]  0.7071068  -0.7071068
[2,]  0.7071068   0.7071068

# finally, lets display our data.
# plotted with the ellipse that best fits the data

ggplot(datum, aes(x = x, y = y)) +
   geom_point() +
   stat_conf_ellipse(bary = T, col = "black")

# the output is displayed below
```

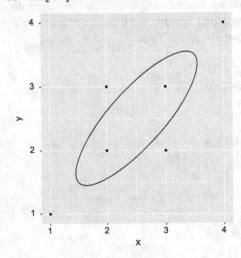

A.2 Persistent Homology via `scikit-tda`

There are a number of packages available for TDA; the landscape is rapidly evolving and different packages have different strengths. For example

- JavaPlex: plays well with `Matlab` and `java`,
      http://appliedtopology.github.io/javaplex
- Ripser: superfast computation of Rips complexes
      https://github.com/Ripser/ripser

We illustrate the `scikit-tda` package, which is a python based public domain package for TDA, available at https://scikit-tda.org/libraries.html

```
# these commands get us set up to plot
import numpy as np
from ripser import ripser
from persim import plot_diagrams
import matplotlib.pyplot as plt


# our first example grabs 100 random points in the plane
data = np.random.random((100,2))
plt.scatter(data[:,0],data[:,1])
plt.show()
```

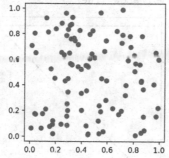

```
# and then plot the persistence diagram
diagrams = ripser(data)['dgms']
plot_diagrams(diagrams, show=True)
```

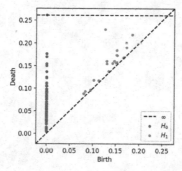

```
# more interesting data set
# scale all points in our random data set to norm 1.

data = data-0.5
data = data/np.linalg.norm(data,ord=2,axis=1).reshape((-1,1))
plt.scatter(data[:,0],data[:,1])
plt.show()
```

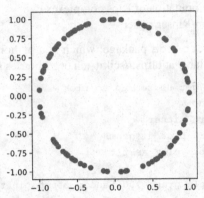

```
# and then plot the persistence diagram
diagrams = ripser(data)['dgms']
plot_diagrams(diagrams, show=True)
```

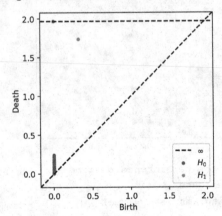

```
# now add some noise to the nice circle data
ddata = np.random.random((10,2))-0.5

ddata=np.vstack([data,ddata])
plt.scatter(ddata[:,0],ddata[:,1])
plt.show()
```

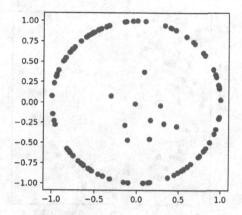

```
# and then plot the persistence diagram
diagrams = ripser(dddata)['dgms']

plot_diagrams(diagrams, show=True)
```

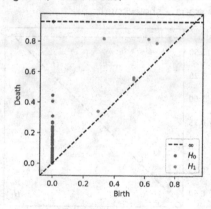

```
# instead of random points, add noise to the points of the circle

rng=np.random.default_rng()
data=data+rng.normal(scale=0.1,size=data.shape)
plt.scatter(data[:,0],data[:,1])
plt.show()
```

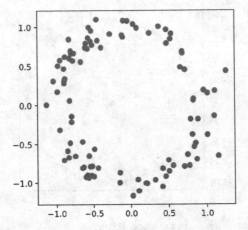

```
# and then plot the persistence diagram

diagrams = ripser(data)['dgms']
plot_diagrams(diagrams, show=True)
```

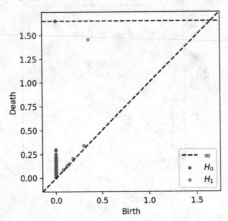

A.3 Computational Algebra via `Macaulay2`

There are many public domain packages for computations in algebra, such as `Sage`, `Singular`, `Macaulay2`. We give snapshots of computations using `Macaulay2`, available at http://www2.macaulay2.com/Macaulay2/.

In Example 8.3.18, the object of interest is a $\mathbb{Z}^3$ graded chain complex

$$0 \longrightarrow R^4 \xrightarrow{d_2} R^6 \xrightarrow{d_1} R^4 \longrightarrow 0.$$

```
i1 : --First, create the ring R and the differentials d1 and d2.
     R=ZZ/101[x_1,x_2,x_3, Degrees=>{{1,0,0},{0,1,0},{0,0,1}}];

i2 : d2 = matrix{{ x_2,x_1,0,0},{-1,0,1,0},{0,-x_1,-x_2,0},{x_3,0,0,x_1},{0,x_3,0,-x_2}, {0,0,x_3,x_1}};

              6      4
o2 : Matrix R  <--- R

i3 : d1 = matrix {{-x_3,-x_2*x_3,-x_3,0,0,0},{x_3,0,0,-x_2,-x_1,0},{0,x_2*x_3,0,x_2,0,-x_2},{0,0,x_3,0,x_1,x_2}};

              4      6
o3 : Matrix R  <--- R

i4 : --Then we have to adjust for the grading
     assert(target d2 != source d1)

i5 : d2' = map(source d1, , d2)

o5 = {0, 0, 1} | x_2  x_1   0     0    |
     {0, 1, 1} | -1   0     1     0    |
     {0, 0, 1} | 0    -x_1  -x_2  0    |
     {0, 1, 0} | x_3  0     0     x_1  |
     {1, 0, 0} | 0    x_3   0     -x_2 |
     {0, 1, 0} | 0    0     x_3   x_1  |

              6      4
o5 : Matrix R  <--- R

i6 : assert(target d2' === source d1)

i7 : --build the chaing complex, and compute the homology. The HH_i command
     --computes the i^th homology, and prune gives a minimal presentation.
     CDelta = chainComplex{d1,d2'}

        4      6      4
o7 = R  <--- R  <--- R

        0      1      2

o7 : ChainComplex

i8 : HH_2(CDelta)

o8 = image {0, 1, 1} | x_1  |
           {1, 0, 1} | -x_2 |
           {0, 1, 1} | x_1  |
           {1, 1, 0} | -x_3 |

                             4
o8 : R-module, submodule of R

i9 : prune HH_2(CDelta)

        1
o9 = R

o9 : R-module, free, degrees {{1, 1, 1}}

i10 : HH_1(CDelta)

o10 = subquotient ({0, 0, 1} | x_2 -x_1 0     0    |, {0, 0, 1} | x_2 x_1   0     0    |)
                   {0, 1, 1} | -1  0    -1    0    |  {0, 1, 1} | -1  0     1     0    |
                   {0, 0, 1} | 0   x_1  x_2   0    |  {0, 0, 1} | 0   -x_1 -x_2   0    |
                   {0, 1, 0} | x_3 0    0     x_1  |  {0, 1, 0} | x_3 0     0     x_1  |
                   {1, 0, 0} | 0   -x_3 0     -x_2 |  {1, 0, 0} | 0   x_3   0     -x_2 |
                   {0, 1, 0} | 0   0    -x_3  x_1  |  {0, 1, 0} | 0   0     x_3   x_1  |

                                6
o10 : R-module, subquotient of R

i11 : prune HH_1(CDelta)

o11 = 0
```

In Example 8.4.2, we computed certain Ext modules and associated primes for

$$\langle x_1^2, x_1 x_2 \rangle \subseteq R = \mathbb{K}[x_1, x_2].$$

```
i1 : R=ZZ/101[x_1,x_2];

i2 : I = matrix{{x_1^2,x_1*x_2}};

             1        2
o2 : Matrix R  <--- R

i3 : M= coker I;

i4 : rM=res(M)

        1      2      1
o4 = R  <-- R  <-- R  <-- 0

        0      1      2      3

o4 : ChainComplex

i5 : rM.dd

           1                            2
o5 = 0 : R  <--------------------- R  : 1
                  | x_1^2 x_1x_2 |

           2                   1
       1 : R  <----------------- R  : 2
                   {2} | -x_2 |
                   {2} | x_1  |

           1
       2 : R  <----- 0 : 3
                 0

o5 : ChainComplexMap

i6 : Ext^2(M,R)

o6 = cokernel {-3} | x_2 x_1 |

                                  1
o6 : R-module, quotient of R

i7 : Ext^1(M,R)

o7 = cokernel {-1} | x_1 |

                                 1
o7 : R-module, quotient of R

i8 : Ext^0(M,R)

o8 = image 0
```

A.4 Multiparameter Persistence via RIVET

The RIVET package provides for computation of 2D MPH, and is available at
https://rivet.readthedocs.io/en/latest/index.html.

RIVET provides visualizations for 2D persistent homology, and has been used
recently to study aspects of tumor evolution in [146]. If a bifiltration is already
in hand RIVET allows the analysis starting from that point. For the bigraded
filtration of Example 8.1.4, relabelling vertices $\{a, \ldots, f\}$ as $\{0, \ldots, 5\}$, we input it
in RIVET syntax as

```
--datatype bifiltration
--xlabel time of appearance
--ylabel network distance

#data
0 ; 1 1
1 ; 0 0
2 ; 0 0
3 ; 1 0
4 ; 0 0
5 ; 0 0
2 4 ; 0 1
2 3 ; 1 0
3 4 ; 1 0
4 5 ; 2 0
1 5 ; 0 2
0 1 ; 1 2
0 5 ; 1 2
1 2 ; 2 2
0 1 5 ; 2 2
2 3 4 ; 3 1
```

For H_0 we obtain the visualization below. The Hilbert function is coded via
greyscale: in the light grey region northeast of $(2, 2)$ we have $HF = 1$ (for the
free summand $R(-2, -2)$); the dark grey in the southwest has $HF = 4$.

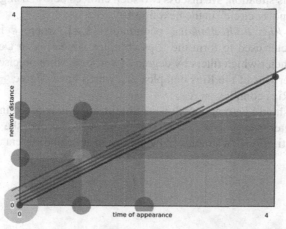

The light dots give degrees where generators of H_0 appear, and the dark dots indicate degrees of relations; size of the dot scales depending on the number of generators in that degree. RIVET allows the user to fix the slope of a line (shown in blue), and displays the barcode of the MPH module when restricted to the line. In Example 8.1.4 there are four generators for H_0 appearing in bidegree $(0, 0)$, and these restrict to give the four purple diagonal lines starting in the southwest corner of the previous figure. In the quadrant to the northeast of position $(3, 2)$ there is only one purple line, reflecting that H_0 has a single connected component in that region.

Continuing with Example 8.1.4, for H_1, RIVET gives the visualization below. The green dot at position $(1, 1)$ heralds the birth of a class in H_1, and a second class is born at $(1, 2)$. At position $(2, 2)$, there is a birth (the four cycle), but the class born at $(1, 2)$ dies; this is reflected by the middle dot, obtained from superposition of a dark dot and a light dot. Finally, there is a death at position $(3, 1)$.

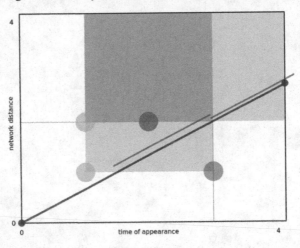

It is somewhat atypical to begin with a bifiltration; a more common starting point is from point cloud data X, and RIVET provides several choices for producing a bifiltration in this situation. We discuss two such choices below, using as underlying data set X the "noisy circle" in the next figure.

The *Degree-Rips Bifiltration* has parameters $\{\delta, \epsilon\}$, where ϵ is the "usual" distance parameter used to form the Rips complex appearing in Definition 4.4.8, and δ is a parameter which filters by degree of vertices. More precisely, let $X_{\delta,\epsilon}$ be subset of the vertices of the Rips complex $\mathscr{R}_\epsilon$ which have degree $\geq \delta$, and define $\mathscr{R}_{\delta,\epsilon}$ to be the Rips complex on $X_{\delta,\epsilon}$.

Two things should be mentioned: first, the filtration is indexed in reverse order in the δ parameter: $\mathscr{R}_{\delta',\epsilon'} \subseteq \mathscr{R}_{\delta,\epsilon}$ if $\epsilon' \leq \epsilon$ and $\delta \leq \delta'$. Second, the corresponding filtration is multi-critical: a simplex may be born in several different degrees.

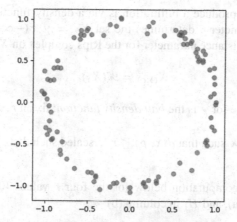

For the noisy circle above, using the Degree-Rips Bifiltration, RIVET gives the results below for H_0 (top figure) and H_1 (bottom figure)

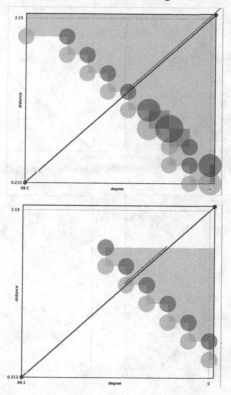

Note that at a fixed degree δ, when there is a birth at distance ϵ' it is often followed by a death at a larger ϵ value. To understand this consider the Rips complex $\mathscr{R}_1$ in Example 4.4.10. The square generates a homology class, and as the distance parameter increases the (empty) square "fills in", and the class in homology perishes.

Another way to produce a bifiltration is via a density function. For a map $\gamma :$ $X \to \mathbb{R}$, the parameter δ determines the set $X_\delta = \gamma^{-1}(-\infty, \delta] \subseteq X$, and the parameter ϵ is the distance parameter for the Rips complex on X_δ, that is

$$X_{(\delta, \epsilon)} = \mathscr{R}(X_\delta)_\epsilon.$$

One common choice for γ is the *ball density function*: fix an r, and let

$$\gamma(x) = |\ p \in X \text{ such that } d(x, p) \leq r\ |, \text{ scaled such that } \sum_{x \in X} \gamma(x) = 1.$$

We carry out this computation below for the four r values $\{0, .25, 1, 10\}$. H_0 is plotted in column (a), and H_1 in column (b)

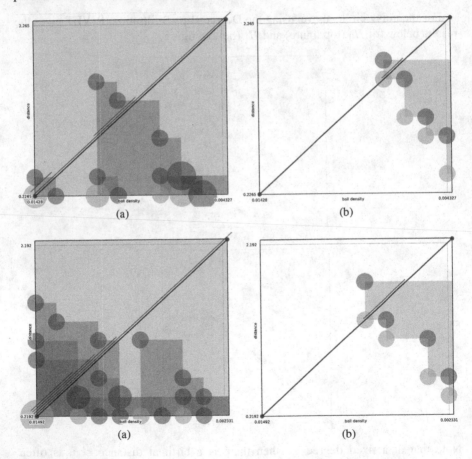

(a)   (b)

(a)   (b)

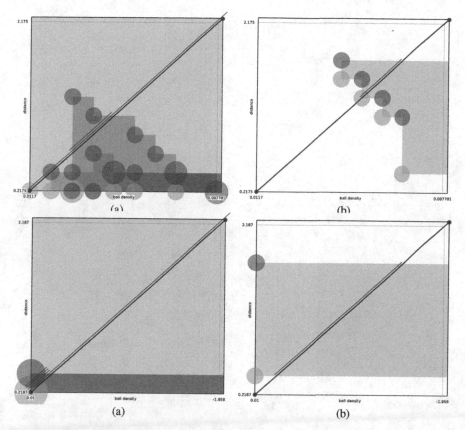

(a)   (b)

(a)   (b)

The RIVET input screen for $r = 1$ is shown below:

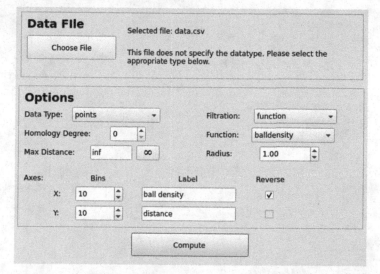

Bibliography

1. H. Adams, G. Carlsson, Evasion paths in mobile sensor networks, *International Journal of Robotics Research*, **34**, 90-104 (2014).
2. P. Aluffi, *Algebra: Chapter 0*, Graduate Studies in Mathematics, 104. American Mathematical Society, Providence, 2009.
3. M. Artin, *Algebra*, Prentice Hall, Inc., Englewood Cliffs, NJ, 1991.
4. M. Atiyah, I. G. MacDonald, *Introduction to Commutative Algebra*, Addison-Wesley, Reading, MA, 1969.
5. D. Attali, H. Edelsbrunner, Y. Mileyko, Weak witnesses for Delaunay triangulations of submanifolds, *Proc. 2007 ACM symposium on solid and physical modeling*, 143-150, (2007).
6. U. Bauer, H. Edelsbrunner, The Morse theory of Čech and Delaunay complexes, *Trans. Amer. Math. Soc.*, **369**, 3741-3762 (2017).
7. U. Bauer, M. Lesnick, Induced matchings and the algebraic stability of persistence barcodes, *J. Comput. Geom.* **6**, 162-191 (2015).
8. U. Bauer, M. Lesnick, Persistence Diagrams as Diagrams: A Categorification of the Stability Theorem, *Topological Data Analysis: the Abel Symposium 2018*, Springer, 67-96 (2020).
9. P. Bendich, H. Edelsbrunner, D. Morozov, A. Patel, Homology and robustness of level and interlevel sets, *Homology Homotopy Appl.*, **15**, 51-72 (2013).
10. P. Bendich, J.S. Marron, E. Miller, A. Pieloch, S. Skwerer, Persistent homology analysis of brain artery trees, *Ann. Appl. Stat.*, **10**, 198-218 (2016).
11. S. Biasotti, A. Cerri, P. Frosini, D. Giorgi, C. Landi, Multidimensional size functions for shape comparison, *Journal of Mathematical Imaging and Vision* **32**, 161-179 (2008).
12. H. Bjerkevik, On the Stability of Interval Decomposable Persistence Modules, *Discrete and Computational Geometry*, **66**, 92-121 (2021).
13. H. Bjerkevik, M. Botnan, Computational complexity of the interleaving distance, *34th International Symposium on Computational Geometry*, Leibniz Int. Proc. Inform., **99**, (2018).
14. H. Bjerkevik, M. Botnan, M. Kerber, Computing the Interleaving Distance is NP-hard, *Foundations of Computational Mathematics*, **20**, 1237-1271 (2020).
15. H. Bjerkevik, M. Lesnick, ℓ^p-Distances on Multiparameter Persistence Modules, preprint, arXiv:2106.13589, (2021).
16. A. Björner, Nerves, fibers and homotopy groups, *J. Combin. Theory Ser. A*, **102**, 88-93 (2003).
17. A. Björner, M. Tanzer, Combinatorial Alexander duality—a short and elementary proof, *Discrete Comput. Geom.*, **42**, 586-593 (2009).
18. M. Botnan, M. Lesnick, Algebraic stability of zigzag persistence modules, *Algebr. Geom. Topol.*, **18**, 3133-3204 (2018).
19. M. Botnan, M. Lesnick, An Introduction to Multiparameter Persistence, *Proceedings of the 2020 International Conference on Representations of Algebras*, to appear.

20. R. Bott, Morse theory indomitable, *Inst. Hautes Études Sci. Publ. Math*, **68**, 99-114, (1989).
21. R. Bott, L. Tu, *Differential forms in algebraic topology*, Springer-Verlag, New York-Berlin (1982)
22. P. Bubenik, V. de Silva, J. Scott, Metrics for generalized persistence modules, *Foundations of Comp. Math.*, **15**, 1501-1531 (2015).
23. P. Bubenik, J. Scott, Categorification of persistent homology, *Discrete and Computational Geometry* **51**, 600-627 (2014).
24. H. Byrne, H. Harrington, R. Muschel, G. Reinert, B. Stolz, U. Tillmann, Topology characterises tumour vasculature, *Math. Today (Southend-on-Sea)*, **55**, 206-210 (2019).
25. G. Carlsson, Topology and data, *Bulletin of the American Mathematical Society*, **46**, 255-308 (2009).
26. G. Carlsson, V. de Silva, ZigZag Persistence, *Found. Comput. Math.*, **10**, 367-405 (2010).
27. G. Carlsson, V. de Silva, S. Kališnik, D. Morozov, Parametrized homology via zigzag persistence, *Algebr. Geom. Topol.*, **19**, 657-700 (2019).
28. G. Carlsson, A. Levine, M. Nicolau, Topology based data analysis identifies a subgroup of breast cancers with a unique mutational profile and excellent survival, *PNAS*, **108** (2011), 7265-7270.
29. G. Carlsson, G. Singh, A. Zomorodian, Computing multidimensional persistence, *Journal of Computational Geometry*, **1**, 72-100 (2010).
30. G. Carlsson, M. Vejdemo-Johansson, *Topological Data Analysis with Applications*, Cambridge University Press, (2022).
31. G. Carlsson, J. Skryzalin, Numeric invariants from multidimensional persistence, *J. Appl. Comput. Topol.*, **1**, 89-119 (2017).
32. G. Carlsson, A. Zomorodian, Computing persistent homology, *Discrete Comput. Geom.*, **33**, 249-274 (2005).
33. G. Carlsson, A. Zomorodian, The theory of multidimensional persistence, *Discrete & Computational Geometry*, **42**, 71-93 (2009).
34. M. Carrière, A. Blumberg, Multiparameter Persistence Images for Topological Machine Learning, *Advances in Neural Information Processing Systems 33 (NeurIPS 2020)*, 13 pp, (2020).
35. W. Chachólski, O. Gäfvert, Stable Invariants for Multidimensional Persistence, *in preparation*.
36. W. Chachólski, M. Scolamiero, and F. Vaccarino, Combinatorial presentation of multidimensional persistent homology, *Journal of Pure and Applied Algebra*, **221**, 1055-1075 (2017).
37. J. Chan, G. Carlsson, R. Rabadan, Topology of viral evolution, *PNAS*, **110**, 18566-18571 (2013).
38. R. Charney, M. Davis, The Euler Chracteristic of a nonpositively curved, piecewise Euclidean manifold, *Pacific J. Math.* **171**, 117-137 (1995).
39. F. Chazal, D. Cohen-Steiner, M. Glisse, L. Guibas, S. Oudot, Proximity of persistence modules and their diagrams, *Proceedings of the 25th Annual Symposium on Computational Geometry, SCG '09*, 237-246 (2009).
40. F. Chazal, V. de Silva, M. Glisse, S. Oudot, The structure and stability of persistence modules, *SpringerBriefs in Mathematics*, Springer, (2016).
41. F. Chazal, A. Lieutier, Weak feature size and persistant homology: computing homology of solids in $\mathbb{R}^n$ from noisy data samples, *Computational geometry (SCG'05)*, ACM, New York, 255-262 (2005).
42. F. Chazal, A. Lieutier, Topology guaranteeing manifold reconstruction using distance function to noisy data, *Computational geometry (SCG'06)*, ACM, New York, 112-118 (2006).
43. F. Chazal, S. Oudot, Towards persistence-based reconstruction in Euclidean spaces, *Computational geometry (SCG'08)*, ACM, New York, 232-241 (2008).
44. D. Cohen-Steiner, H. Edelsbrunner, J. Harer, Stability of persistence diagrams, *Discrete Comput. Geom.* **37**, 103-120 (2007).
45. D. Cohen-Steiner, H. Edelsbrunner, J. Harer, Extending persistence using Poincaré and Lefschetz duality, *Found. Comput. Math.*, **9**, 79-103 (2009).

46. D. Cohen-Steiner, H. Edelsbrunner, D. Morozov, Vines and Vineyards by Updating Persistence in Linear Time, *Proceedings of the Annual ACM Symposium on Computational Geometry*, 119-126 (2006).
47. D. Cox, J. Little, D. O'Shea, *Ideals, Varieties and Algorithms*, third edition, Undergraduate Texts in Math., Springer, New York, 2007.
48. D. Cox, J. Little, H. Schenck, *Toric Varieties*, AMS Graduate Studies in Math, AMS, (2010).
49. M. Davis, B. Okun, Vanishing theorems and conjectures for the l_2- homology of right-angled Coxeter groups, *Geom. Topo.*, **5**, 7-74 (2001).
50. M. DeGroot, Reaching a consensus, *Journal of the American Statistical Association*, **69**, 118-121 (1974).
51. V. de Silva, G. Carlsson, Topological estimation using witness complexes, *Proc. 1st Eurographics conference on Point-Based Graphics*, 157-166 (2004).
52. V. de Silva, R. Ghrist, Coordinate-free Coverage in Sensor Networks with Controlled Boundaries via Homology, *International Journal of Robotics Research*, **25**, 1205-1222 (2006).
53. V. de Silva, R. Ghrist, Coverage in sensor networks via persistent homology, *Algebr. Geom. Topol.* **7**, 339-358 (2007).
54. V. de Silva, R. Ghrist, Homological Sensor Networks, *Notices of the A.M.S.*, **54**, 10-17 (2007).
55. V. de Silva, D. Morozov, M. Vejdemo-Johansson, Dualities in Persistent Homology, *Inverse Problems*, **27**, 17pp, (2011).
56. V. de Silva, E. Munch, A. Patel, Categorified Reeb graphs, *Discrete Comput. Geom.*,**55**, 854-906 (2016).
57. V. de Silva, E. Munch, A. Stefanou, Theory of interleavings on categories with a flow, *Theory Appl. Categ.* **33**, 583-607 (2018).
58. T. Dey, F. Fan, Y. Wang, Computing topological persistence for simplicial maps, *Computational geometry (SoCG'14)*, ACM, New York, 345-354 (2014).
59. T. Dey, T. Li, Y. Wang, An efficient algorithm for 1-dimensional persistent path homology, *36th International Symposium on Computational Geometry, Leibniz Int. Proc. Inform*, **136**, 15pp (2019).
60. T. Dey, Y. Wang, *Computational Topology for Data Analysis*, Cambridge University Press, (2022).
61. B. Di Fabio, C. Landi, Reeb graphs of piecewise linear functions, *Graph-based representations in pattern recognition* Lecture Notes in Comput. Sci., **10310**, 23-35 (2017).
62. B. Di Fabio, C. Landi, The edit distance for Reeb graphs of surfaces, *Discrete Comput. Geom.*, **55**, 423-461 (2016).
63. B. Di Fabio, C. Landi, Reeb graphs of curves are stable under function perturbations, *Math. Methods Appl. Sci.*, **35**, 1456-1471 (2012).
64. H. Edelsbrunner, D. Kirkpatrick, R. Seidel, *On the shape of a set of points in the plane*, IEEE Trans. Info. Theory, **29**, 551-559 (1983).
65. H. Edelsbrunner, D. Letscher, A. Zomorodian, *Topological persistence and simplification*, Discrete Comput. Geom., **28**, 511-533 (2002).
66. H. Edelsbrunner, J. Harer, Persistent homology — A survey, *Contemporary Mathematics*, **453**, 257-282 (2008).
67. H. Edelsbrunner, J. Harer, *Computational Topology: An Introduction*, American Mathematical Society, Providence (2010).
68. H. Edelsbrunner, M. Kerber, Alexander duality for functions: the persistent behavior of land and water and shore, *Computational geometry (SCG'12)*, 249-258, (2012).
69. H. Edelsbrunner, D. Morozov, Persistent homology: theory and practice, *European Congress of Mathematics*, Eur. Math. Soc., Zürich, 31-50 (2013).
70. D. Eisenbud, *Commutative Algebra with a View Toward Algebraic Geometry*, Graduate Texts in Math. **150**, Springer, New York, 2000.
71. D. Eisenbud, C. Huneke, W. Vasconcelos, Direct methods for primary decomposition, *Invent. Math.*, **110**, 207-235 (1992).

72. N. Friedkin, E. Johnsen, Social influence and opinions, *J. Mathematical Sociology*, **15**, 193-206 (1990).

73. P. Frosini, A distance for similarity classes of submanifolds of a Euclidean space, *Bulletin of the Australian Mathematical Society*, **42**, 407-415 (1990).

74. P. Frosini, Discrete computation of size functions. *J. Combin. Inform. System Sci.*, **17**, 132-250 (1992).

75. P. Frosini, C. Landi, Size functions and morphological transformations, *Acta Appl. Math.* **49**, 85-104 (1997).

76. W. Fulton, *Algebraic Topology-A first course*, Graduate Texts in Math. **153**, Springer, New York, 1995.

77. P. Gabriel, Unzerlegbare darstellung I, *Manuscripta Math.*, **6**, 71-103 (1972).

78. R. Ghrist, Barcodes: the persistent topology of data, *Bull. Amer. Math. Soc.*, **45**, 61-75 (2008).

79. R. Ghrist, *Elementary Applied Topology*, ed. 1.0, Createspace, (2014).

80. R. Ghrist, *Homological algebra and data*, The mathematics of data, p. 273-325, IAS/Park City Math. Ser., **25**, Amer. Math. Soc., Providence, 2018.

81. R. Ghrist, Y. Hiraoka, Applications of sheaf cohomology and exact sequences to network coding, *Proc. NOLTA*, 2011.

82. R. Ghrist, S. Krishnan, Positive Alexander duality for pursuit and evasion, *SIAM J. Appl. Algebra Geom.*, **1**, 308-327 (2017).

83. M. Greenberg, J. Harper, *Algebraic Topology: A First Course*, Benjamin/Cummings, Reading, MA, 1981. Reprinted by Westview Press, Boulder, CO.

84. P. Griffiths, J. Harris, *Principles of algebraic geometry*, Pure and Applied Mathematics. John Wiley & Sons, New York, 1978.

85. L. Guibas, S. Oudot, Reconstruction using witness complexes, *Discrete and Computational Geometry*, **40**, 325-356 (2008).

86. J. Hansen, R. Ghrist, Toward a spectral theory of cellular sheaves, *J. Appl. Comput. Topol.*, **3**, 315-358 (2019).

87. J. Hansen, R. Ghrist, Opinion dynamics on discourse sheaves, *SIAM J. Appl. Math.*,**81**, 2033-2060 (2021).

88. H. Harrington, N. Otter, H. Schenck, U. Tillmann, Stratifying multiparameter persistent homology, *SIAM J. Appl. Algebra Geom.*, **3**, 439-471 (2019).

89. J. Harris, *Algebraic Geometry: A First Course*, Graduate Texts in Math. **133**, Springer, New York, 1992.

90. R. Hartshorne, *Algebraic Geometry*, Graduate Texts in Math. **52**, Springer, New York, 1977.

91. A. Hatcher, *Algebraic Topology*, Cambridge Univ. Press, Cambridge, 2002.

92. Y. H. He, *The Calabi-Yau landscape: from geometry to physics to machine learning*, Springer, New York, 2021.

93. T. Hungerford, *Algebra*, Graduate Texts in Math. **73**, Springer, New York, 2002.

94. X. Jiang, L.-H. Lim, Y. Yao, Y. Ye, Statistical ranking and combinatorial Hodge theory, *Math. Program.*, **127**, 203-244 (2011).

95. S. Kališnik, Alexander duality for parametrized homology, *Homology Homotopy Appl.*, **15**, 227-243 (2013).

96. S. Kališnik, Tropical coordinates on the space of persistence barcodes, *Found. Comput. Math.*, **19**, 101-129 (2019).

97. S. Kališnik, C. Lehn, V. Limic, Geometric and probabilistic limit theorems in topological data analysis, *Adv. in Appl. Math.*, **131**, 36pp, (2021).

98. L. Kanari, A. Garin, K. Hess, From trees to barcodes and back again: theoretical and statistical perspectives, *Algorithms (Basel)*, **13**, 27pp (2020).

99. M. Kerber, M. Lesnick, S. Oudot, Exact computation of the matching distance on 2-parameter persistence modules, *J. Comput. Geom.*, **11**, 4-25 (2020).

100. K. Knudson, A refinement of multidimensional persistence, *Homology, Homotopy and Applications*, **10**, 259-281 (2008).

101. D. Kozlov, *Combinatorial algebraic topology*, Algorithms and Computation in Mathematics, 21. Springer, Berlin (2008).

102. S. Lang, *Algebra*, Graduate Texts in Math, **211** Springer, New York, 2002.
103. M. Lesnick, The theory of the interleaving distance on multidimensional persistence modules, *Found. Comput. Math.*, **15** 613-650 (2015).
104. R. Lewis, Spectral Sequences and Applied Topology, Ph.D. thesis, Stanford (2016).
105. R. Lewis, D. Morozov, Parallel computation of persistent homology using the blowup complex *Annual ACM Symposium on Parallelism in Algorithms and Architectures*, 323-331, (2015).
106. L. H. Lim, Hodge Laplacians on Graphs, *SIAM Rev.*, **62**, 685-715, (2020).
107. S. MacLane, *Categories for the working mathematician*, Graduate Texts in Mathematics, Vol. 5. Springer-Verlag, New York-Berlin, 1971.
108. Y. Matsumoto, *An Introduction to Morse Theory*, American Mathematical Society, Providence, RI, 2002.
109. J. McCleary, *A User's Guide to Spectral Sequences*, Cambridge Univ. Press, Cambridge, 2001.
110. E. Miller, Fruit flies and moduli: interactions between biology and mathematics, *Notices Amer. Math. Soc.*, **62**, 1178-1184 (2015).
111. J. Milnor, *Morse Theory*, Princeton University Press, 1963
112. N. Milosavljević, D. Morozov, P. Škraba, Zigzag persistent homology in matrix multiplication time, *Computational geometry (SCG'11)*, 216-225, (2011).
113. D. Morozov, Persistence algorithm takes cubic time in worst case, *Proceedings of Topology-Based Methods in Visualization*, (2013).
114. D. Morozov, K. Beketayev, G. Weber, Interleaving Distance between Merge Trees, *Proceedings of Topology-Based Methods in Visualization*, (2013).
115. M. Morse, Relations between the critical points of a real function of n independent variables, *Trans. Amer. Math. Soc.*, **27**, 345-396 (1925).
116. M. Morse, The foundations of a theory in the calculus of variations in the large, *Trans. Amer. Math. Soc.* **30**, 213-274 (1928).
117. M. Morse, Rank and span in functional topology, *Ann. of Math*, **41**, 419-454 (1940).
118. E. Munch, B. Wang, Convergence between categorical representations of Reeb space and mapper, *32nd International Symposium on Computational Geometry, Leibniz Int. Proc. Inform.*, **34**, 461-475 (2016),
119. J. Munkres, *Elements of Algebraic Topology*, Addison-Wesley, Reading, MA, 1984. Reprinted by Westview Press, Boulder, CO.
120. B. Nelson, Parameterized Topological Data Analysis, Ph.D. thesis, Stanford (2020).
121. P. Niyogi, S. Smale, S. Weinberger, Finding the homology of submanifolds with high confidence from random samples, *Discrete Comput. Geom.*, **39**, 419-441 (2008).
122. P. Oesterling, C. Heine, G. Weber, D. Morozov, G. Scheuermann, Computing and visualizing time-varying merge trees for high-dimensional data, *Topological methods in data analysis and visualization, IV*, Math. Vis., Springer, 87-101 (2017).
123. N. Otter, M. Porter, U. Tillmann, P. Grindrod, H. Harrington, A roadmap for the computation of persistent homology, *EJP Data Science*, **6**, 1-38 (2017).
124. S. Oudot, *Persistence theory: from quiver representations to data analysis*, American Mathematical Society, Providence, RI, 2015.
125. J. Perea, A brief history of persistence, *Morfismos*, **23**, 1-16 (2019).
126. L. Polterovich, D. Rosen, K. Samvelyan, J. Zhang, *Topological Persistence in Geometry and Analysis*, AMS University Lecture Series, AMS, 2020.
127. R. Rabadan, A. Blumberg, *Topological Data Analysis for Genomics and Evolution: Topology in Biology*, Cambridge University Press, 2020.
128. E. Riehl, *Category Theory in Context*, Dover, 2016.
129. V. Robins, Towards computing homology from finite approximations, *Topology Proceedings*, **24**, 503-532 (1999).
130. V. Robins, K. Turner, Principal component analysis of persistent homology rank functions with case studies of spatial point patterns, sphere packing and colloids, *Phys. D*, **334**, 99-117 (2016).

131. M. Robinson, *Topological signal processing*, Springer, Heidelberg, 2014.

132. J. J. Rotman, *Advanced Modern Algebra*, Prentice-Hall, Upper Saddle River, NJ, 2002.

133. H. Schenck, *Computational Algebraic Geometry*, Cambridge Univ. Press, Cambridge, 2003.

134. M. Scolamiero, W. Chachólski, A. Lundman, R. Ramanujam, S. Öberg, Multidimensional Persistence and Noise, *Foundations of Computational Mathematics*, **17**, 1367-1406 (2017).

135. I. Singer, J. Thorpe, *Lecture notes on elementary topology and geometry*, Undergraduate Texts in Math., Springer, New York, 1967.

136. G. Singh, F. Mémoli, T. Ishkhanov, G. Sapiro, G. Carlsson, D. Ringach, Topological analysis of population activity in visual cortex, *Journal of Vision*, **8**, 1-18 (2008).

137. E. Spanier, *Algebraic Topology*, McGraw-Hill, New York, 1966. Reprinted by Springer Verlag, New York.

138. M. Spivak, *Calculus on Manifolds*, W.A. Benjamin, New York, 1965.

139. J. Stewart, *Calculus*, Cengage, 2020.

140. B. Stolz, H. Harrington, M. Porter, Persistent homology of time-dependent functional networks constructed from coupled time series, *Chaos* **27**, 17pp, (2017).

141. B. Stolz, J. Tanner, H. Harrington, V. Nanda, Geometric anomaly detection in data, *Proceedings of the National Academy of Sciences*, **117**, 19664-19669, (2021).

142. M. Taylor, Towards a mathematical theory of influence and attitude change, *Human Relations*, **21**, 121-139 (1968).

143. K. Turner, Medians of populations of persistence diagrams, *Homology Homotopy Appl.*, **22**, 255-282 (2020).

144. K. Turner, S. Mukherjee, D. Boyer, Persistent homology transform for modeling shapes and surfaces, *Information and Inference*, **3**, 310-344 (2014).

145. K. Turner, Y. Mileyko, S. Mukherjee, J. Harer, Frechet means for distributions of persistence diagrams, *Discrete Comput. Geom.*, **52**, 44-70 (2014).

146. O. Vipond, J. Bull, P. Macklin, U. Tillman, C. Pugh, H. Byrne, H. Harrington, Multiparameter persistent homology landscapes identify immune cell spatial patterns in tumors, *Proceedings of the National Academy of Sciences*, **118**, 9pp, (2021).

147. C. Voisin, *Hodge theory and complex algebraic geometry I, II*, Cambridge Studies in Advanced Mathematics, 76. Cambridge University Press, Cambridge, 2007.

148. C. Webb, Decomposition of graded modules, *Proceedings of the American Mathematical Society*, **94**, 565-571 (1985).

149. C. Weibel, *An introduction to homological algebra*, Cambridge Univ. Press, Cambridge, 1994.

150. S. Weinberger, What is... persistent homology?, *Notices Amer. Math. Soc.* **58**, 36-39 (2011).

151. M. Wright, X. Zheng, Topological data analysis on simple English Wikipedia articles, *PUMP J. Undergrad. Res.*, **3**, 308-328 (2020).

152. A. Zomorodian, *Computing and Comprehending Topology: Persistence and Hierarchical Morse Complexes*, Ph.D. thesis, University of Illinois at Urbana-Champaign, 2001.

153. A. Zomorodian, *Topology for computing*, Cambridge University Press, Cambridge, 2005.

Index

Printed in the United States
by Baker & Taylor Publisher Services